Nonlinear Tools: Theory Meets Diverse Problems

Chopra

Contents

Introduction

In this dissertation, we explore several topics within the general area of *nonlinear algebra*. Nonlinear algebra revolves around algebraic methods based on systems of mutivariate polynomial equations, with an emphasis both on theoretical results, explicit computational methods, and connections to various applications, like optimization, statistics, complexity theory, mathematical biology, and quantum information theory. At the heart of nonlinear algebra lies algebraic geometry, but it also encompasses methods from combinatorics, representation theory, multilinear algebra, commutative algebra, and convex geometry, among others.

Part I of this book deals with *tensors*. Tensors are the natural higher-dimensional generalization of matrices. Just as matrices are basic objects in linear algebra, tensors are fundamental for nonlinear algebra. After giving a gen-eral introduction to tensors in Chapter 1, we explore three different topics, each revolving around a different application of tensors. The corresponding chapters can be read independently.

- Chapter 2 is about *matrix product states*. These are certain varieties of tensors that arise in quantum information theory, but also have connections to numerical optimization. We use methods from nonlinear algebra and algebraic geometry to answer questions about topology, defining equations, and identifiability of uniform matrix product states. In addition, we present a proof of a conjecture in quantum information theory regarding the higher-dimensional version of matrix product states.

- In Chapter 3, we study *fast matrix multiplication*. The problem of determining what is the fastest algorithm for multiplying two matrices is known to be equivalent to estimating the rank of a certain tensor. Two new contributions to the study of fast matrix multiplication are discussed. First, we study the symmetrized matrix multiplication tensor and, using methods from representation theory, gain a better understanding of the symmetry of the space this tensor lives in. Second, we use these results as an inspiration to construct new tensors which could potentially be used to obtain faster algorithms for matrix multiplication.

- In Chapter 4, we present a nice connection between tensors (more precisely: Waring rank), and classical algebraic geometry (more precisely: cubic surfaces). The *Hessian discriminant* is a recently introduced invariant, defined by the locus of cubic surfaces whose equation has Waring rank greater than five. The main result of this chapter is the expression of the Hessian discriminant in terms of fundamental invariants.

Part II of the **book** takes place on the intersection of algebraic geometry and combinatorics. More specifically, we apply algebro-geometric methods to study *matroids* and *flag matroids*. Matroids are one of the central notions in modern combinatorics. They simultaneously generalize the notion of linear independence in a vector space, and the notion of a graph, and have found applications in among others optimization, network theory and coding theory. One of the most important matroid invariants is the Tutte polynomial. We review a geometric interpretation of the Tutte polynomial by Fink and Speyer, and introduce a natural generalization to so-called flag matroids. For matroids, the Tutte polynomial is defined combinatorially, and the K-theoretic interpretation is a property. In contrast, for flag matroids, our K-theoretic description serves as a *definition* for the Tutte polynomial. Obtaining a fully combinatorial description of this *flag-geometric Tutte polynomial* appears to be out of reach. In the final chapter of this **book**, we prove several interesting combinatorial properties of the flag-geometric Tutte polynomial of a flag matroid.

Most of the contents of this **book** have appeared in slightly altered form in several papers. At the moment of writing, two of these papers [Sey18, MSV19] have been published in peer-reviewed journals, two [CDMS20, DS20] have been accepted for publication in 2020, one [CMS19] is under review, and one [DES20] is in preparation.

Algebraic Geometry of Tensors

A d-way tensor is simply an element in a tensor product $V_1 \otimes \cdots \otimes V_d$ of vector spaces $V_1, \ldots, V_d$. After choosing a basis for every V_i, a tensor is given by a d-dimensional table of numbers; in particular, 2-way tensors are just matrices. Tensors provide a useful way for organizing data, so it is no surprise that they appear in a wide variety of applications, like data science, signal processing, phylogenetics, and complexity theory, to name just a few.

A central notion in the theory of tensors is that of tensor rank. We say that a tensor is of *rank one* if it can be written in the form $v_1 \otimes \cdots \otimes v_d$. The *rank* of a tensor T is the smallest integer r such that T can be written as a sum of r rank one tensors. Tensor rank is a straightforward generalization of matrix rank, but for $d > 2$ the notion of rank is much more ill-behaved. For instance, the set of all tensors of rank at most r is not always a closed set. This leads to the notion of *border rank*: the border rank of a tensor T is the smallest r such that T can be arbitrarily closely approximated with tensors of rank r. Or equivalently: the set of tensors of border rank at most r is the (Zariski or Euclidean) closure of the set of tensors of rank at most r. This set can be geometrically described as the r-th secant variety of the Segre embedding.

Many interesting questions can be asked about these secant varieties: What is the dimension? Can we find defining equations? When is the set of tensors of *rank* at most r closed, and can we describe the boundary? When is the parametrizing map generically finite or generically injective? There is an extensive literature studying these varieties [Lan12, Lan17, IK99, Zak05]. But low rank tensors are not the only interesting varieties of tensors. Motivated by applications in among others optimization, quantum information theory, and algebraic statistics, there is an interest in several kinds of tensor formats. One important example of these are tensor network states, and more specifically: matrix product states.

Matrix product states

Matrix product states and uniform matrix product states play a crucial role in quantum physics and quantum chemistry [PGVWC07, SPGWC10, Orú14, Sch11, Hac12, YL18]. They are used, for instance, to compute the eigenstates of the Schrödinger equation. Matrix product states provide a way to represent special tensors in an efficient way and uniform matrix product states are partially symmetric analogs of matrix product states.

In Chapter 2, we introduce new algebraic methods to answer problems and questions coming from this area. In particular, we answer several questions and conjectures posed by Critch, Morton and Hackbusch. Our main emphasis is on interactions among algebraic geometry and matrix product states. While secant varieties and border rank already form a well-established topic within algebraic geometry, tensor networks and matrix product states so far have only received very limited attention (with notable exceptions [LQY12, CM14, PGVWC08]). However, we are sure that this situation is changing. The first four sections of Chapter 2 are based on joint work with Adam Czapliński and Mateusz Michałek [CMS19]; Section 2.5 is based on joint work with Mateusz Michałek and Frank Verstraete [MSV19].

For given parameters (D, d, N), the set of uniform matrix product states is a subset of the tensor space $(\mathbb{C}^d)^{\otimes N}$, defined as the image of a polynomial map

$$T_N : (\mathbb{C}^{D \times D})^d \to (\mathbb{C}^d)^{\otimes N}$$
$$(M_0, \ldots, M_{d-1}) \mapsto \sum_{0 \leq i_1, \ldots, i_N \leq d-1} \mathrm{tr}(M_{i_1} \cdots M_{i_N}) e_{i_1} \otimes \cdots \otimes e_{i_N}.$$

One of the main goals of Chapter 2 is to study geometric and topological properties of the uniform matrix product states $\mathrm{uMPS}(D, d, N)$ and the Zariski closure $\overline{\mathrm{uMPS}(D, d, N)}$, which in case of complex numbers coincides with the Euclidean closure. As an application of our methods we confirm two conjectures of Critch and Morton [CM14]. Precisely, one asserts that, under mild assumptions, for two 2×2 matrices, matrix product states are "identifiable". The main ingredient of our proof is the so-called *fundamental theorem of uniform matrix product states*. We can also verify the conjecture concerning the defining ideal of the closure of the uMPS in special cases. This is related to the probabilistic graphical models known as hidden Markov models and to the conjecture of Bray-Morton-Sturmfels [BM05]. A variant of this conjecture states that for any fixed D and d, the ideal of $\mathrm{uMPS}(D, d, N)$ is generated in low degree for N large enough.

One of the tools we use is the *trace algebra* [Pro76, Sib68]. The applications of trace algebras to the theory of matrix product states were already investigated by Critch and Morton [CM14]. We show how this method can be used to derive the conjectured description of the ideal of $\overline{\mathrm{uMPS}(2, 2, 5)}$. Further, we provide a full description of the ideal of $\overline{\mathrm{uMPS}(2, 2, 6)}$. Moreover, we describe a useful surjectivity criterion for polynomial maps, which can be used to prove that every tensor can be expressed as a uMPS. We apply it to show that $\mathrm{uMPS}(3, 2, 4)$ fills the ambient space of cyclic tensors. Our method is very general and we believe it can be used in many other cases beyond matrix product states.

The very important questions of the closedness of families of tensors that allow representations as matrix product states were asked by W. Hackbusch and L. Grasedyck (cf. [LQY12, HMS19]). One of the questions was, when the sets $\mathrm{uMPS}(D, d, N)$ and $\overline{\mathrm{uMPS}(D, d, N)}$ may differ. We answer the question when $\mathrm{uMPS}(D, d, N)$ is closed, i.e. both sets are equal, in the case $D = 2$. Further, we provide an explicit tensor (the so-called *W-state*) which is always in $\overline{\mathrm{uMPS}(D, d, N)}$, but not in $\mathrm{uMPS}(D, d, N)$ when N is large compared to D, providing many instances where $\mathrm{uMPS}(D, d, N)$ is not closed. Moreover, we study the dimension of $\overline{\mathrm{uMPS}(D, d, N)}$ and the connectedness in a more general set-up. In the table below we collect our results concerning closedness of $\mathrm{uMPS}(D, 2, N)$. The second row is Theorem 2.4.1, the underlined F is Example 2.4.19.

y	x	2	3	4	6	8	14	20
	D\N	1	2	3	4	5	6	7
2	1	F	C	C	C	C	C	C
5	2	F	F	F	N	N	N	N
10	3	F	F	F	$\underline{\underline{\mathrm{F}}}$			
27	4	F	F	F	F			

Table 1: $\mathrm{uMPS}(D, 2, N)$

F: fills the ambient space, C: closed, but does not fill, N: not closed

x: dimension of the ambient space, y: expected dimension

An important result in the study of matrix product states is the *quantum Wielandt theorem*.

Theorem 2.3.7 ([SPGWC10]). *For every $D \in \mathbb{N}$, there exists a constant $C(D)$ such that the following holds:*

> Let $L \subseteq \mathbb{C}^{D \times D}$ be a linear space of $D \times D$-matrices, and assume that
> there is an N such that $L^N = \mathbb{C}^{D \times D}$. Then already for $N = C(D)$, it
> holds that $L^N = \mathbb{C}^{D \times D}$,

were by L^N we mean the linear space spanned by all products of N matrices in L.

The smallest such $C(D)$ is known as the *injectivity radius*. As an example application, the proof that uMPS(D, d, N) is not closed when N is large compared to D requires considering the injectivity radius.

In the final section of Chapter 2, we formulate and prove a generalization of the quantum Wielandt theorem (Theorem 2.5.5). The original quantum Wielandt theorem is a statement about multiplying matrices, whereas our theorem is a statement about contracting $2m$-way tensors arranged in an m-dimensional grid. This "tensor version" of the quantum Wielandt theorem is motivated by *Projected Entangled Pair States* (PEPS), the higher-dimensional generalizations of MPS.

The bounds for quantum Wielandt theorem in [SPGWC10, MS19, Rah20] were obtained using explicit methods from linear algebra. Our main new insight is the application of nonconstructive Noetherian arguments from non-linear algebra. For matrix product states, our theorem proves Conjecture 1 in [PGVWC07].

Above questions are often inspired by applications. However, they may be also seen as analogues of well-studied questions in the theory of secant varieties, rank and border rank. For example, the closedness of the image of a map is a natural question from a theoretical point of view, but also plays an important role in best approximation problems [DSL08, QML19].

Fast matrix multiplication

Another field of research where tensors and tensor rank play a central role is in determining the complexity of matrix multiplication. In 1969, Strassen [Str69] presented his celebrated algorithm for matrix multiplication breaking for the first time the naive complexity bound of n^3 for $n \times n$ matrices. Since then, the complexity of the optimal matrix multiplication algorithm is one of the central problems in computer science. In terms of algebra we know that this question is equivalent to estimating rank or border rank of a specific tensor $M_{\langle n,n,n \rangle} \in \mathbb{C}^{n^2} \otimes \mathbb{C}^{n^2} \otimes \mathbb{C}^{n^2}$ [BCS97, Lan12, Lan17]. The complexity of matrix multiplication is measured by the constant ω, defined as the smallest number such that for any $\epsilon > 0$ the multiplication of $n \times n$ matrices can be performed in time $O(n^{\omega + \epsilon})$. Equivalently, ω is the smallest number such that for any $\epsilon > 0$ the rank (or border rank) of $M_{\langle n,n,n \rangle}$ is $O(n^{\omega + \epsilon})$.

The best known upper bounds on ω are all obtained using the so-called *laser method*, which is based on the work of Strassen [Str87]. The idea behind the laser method is to, instead of studying the matrix multiplication tensor directly, consider a different tensor which can be proven to have low border rank, and at the same time is "close" to being a matrix multiplication tensor in a sense that we will make precise later. Strassen obtained a bound $\omega < 2.48$ using the laser method. Shortly thereafter, Coppersmith and Winograd introduced a new tensor, and applied the laser method to it to obtain $\omega < 2.3755$. Since then, the only improvements on the bound of ω were made by Stothers, Williams, and Le Gall [Sto10, Wil12, LG14], arriving at the current state of the art $\omega < 2.373$. These improvements were all obtained by applying the laser method to the Coppersmith-Winograd tensor. In Section 3.1, we give a brief introduction to the laser method and present a variant, which is slightly more general then the version usually presented in the literature.

Recently, Chiantini et al. [CHI$^+$18] proved that instead of $M_{\langle n,n,n \rangle}$ one can consider the so-called *symmetrized matrix multiplication tensor* SM_n and study its symmetric (border) rank. The polynomial SM_n can be viewed as an element of the SL_n-representation $S^3(\mathfrak{sl}_n^*)$, and is in fact an invariant. This is the motivation for our study of the SL_n-representation $S^k(\mathfrak{sl}_n^*)$, which will be carried out in Section 3.2. The computations of plethysm are in general very hard and explicit formulas are known only in specific cases [Mac98]. For example for symmetric power $S^3(S^k)$ the decomposition was classically computed already in [Thr42, Plu72], but $S^4(S^k)$ and $S^5(S^k)$ were only recently explicitly obtained in [KM16]. As symmetric powers (together with exterior powers) are the simplest Schur functors, one could expect that respective formulas for $S^d(\mathfrak{sl}_n)$ are harder. In principle, one could use the methods of [How87, KM16, MM15] to decompose this plethysm, but this requires a lot of nontrivial character manipulations. Instead, in Section 3.2, we present a very easy proof of explicit decomposition based on Cauchy formula and Littlewood-Richardson rule in Theorem 3.2.3. In fact, using our method one can inductively obtain the formula for $S^k(\mathfrak{sl}_n)$ for any k. While matrix multiplication is represented by the (unique) invariant in $S^3(\mathfrak{sl}_n)$, we also study the other highest weight vectors. Surprisingly, it turns out is that some of the highest weight vectors are variants of the Coppersmith-Winograd tensor.

Recent work [AFLG15, AW18a, AW18b, CVZ19] shows the limitations of the Coppersmith-Winograd tensor, and indicates the need for finding different tensors that are similar to the Coppersmith-Winograd tensor in order to prove better bounds on ω. In Section 3.3, we explore two approaches of constructing such tensors. One approach is based on the above observation about the highest weight vectors in $S^3(\mathfrak{sl}_n)$: since one of them is the Coppersmith-Winograd tensor, we can attempt to use the other highest weight vectors to prove bounds on ω. One of them appears to be of minimal border rank. Assuming this is the case, we can apply the laser method to it and obtain a bound $\omega < 2.451$, which is not as good as the Coppersmith-Winograd bound, but better than Strassen's bound. The other approach builds upon the work of Landsberg–Michałek [LM17] and Bläser–Lysikov [BL16]. They proved that, under certain genericity assumptions, a tensor is of minimal border rank if and only if it is the multiplication tensor of a smoothable finite-dimensional algebra. The Coppersmith-Winograd tensor arises in this way: it is the multiplication tensor of an algebra with Hilbert function $(1, n, 1)$, and it follows from a result of Cartwright et al. [CEVV09] that such an algebra is always smoothable. We give an example of an algebra with Hilbert function $(1, n, 2)$ (hence smoothable by [CEVV09]) whose multiplication tensor is suitable for the laser method. The obtained bound $\omega < 2.431$ is better than the bound above, but still not as good as the bound obtained from the Coppersmith-Winograd tensor. This chapter is based on [Sey18], as well as joint work in progress with Joachim Jelisiejew and Mateusz Michałek.

The Hessian discriminant of a cubic surface.

Symmetric tensors in $\mathrm{Sym}^d(\mathbb{C}^n)$ can be identified with homogeneous polynomials of degree d in n variables. The *Waring rank* (or symmetric rank) of a homogeneous polynomial f is the smallest r for which f can be written as a sum of r powers of linear forms. An important research topic in classical algebraic geometry is the study of cubic surfaces. The rank of a cubic surface is simply the Waring rank of its defining equation; it is known that a general cubic surface has rank 5. Recently, Anna Seigal introduced a new invariant of cubic surfaces called the *Hessian discriminant*. It is a homogeneous degree 120 polynomial in the 20 variables parametrizing the space of cubic surfaces, whose vanishing locus is the Zariski closure of the set of rank 6 cubic surfaces.

Since the Hessian discriminant is invariant under the action of the group $PGL(3)$, it can be expressed in terms of the classically known *fundamental invariants* of cubic surfaces. In Chapter 4, which is based on joint work with Rodica Dinu [DS20], we explain how to do this: we present a proof that the Hessian Discriminant is equal to I_{40}^3, where I_{40} is Salmon's invariant of degree 40. We also present algorithms that allow for explicit computations.

Algebraic Geometry of Matroids

Matroids are one of the central notions in modern combinatorics. They simultaneously generalize the notion of linear independence in a vector space, and the notion of a graph, and have found applications in among others optimization, network theory and coding theory. Given a finite set E of vectors in a vector space, the data of which subsets of E are linearly independent defines a matroid. Matroids that arise in this way are called *representable* or *realizable*. In general, a matroid is defined as a finite set E together with collection of "independent" subsets of E satisfying certain axioms.

The most famous matroid invariant is the *Tutte polynomial*. It was first defined for graphs by Tutte [Tut67] and then for matroids by Crapo [Cra69].

Definition 5.1.11. *Let M be a matroid of rank r on a finite set E. Its* Tutte polynomial $T_M(x, y)$ *is a bivariate polynomial in* x, y *defined by*

$$T_M(x, y) := \sum_{S \subseteq E} (x - 1)^{r(E) - r(S)} (y - 1)^{|S| - r(S)},$$

where the rank $r(S)$ *of a subset* $S \subseteq E$ *is defined as the size of the largest independent subset of* S.

The interplay of matroids and geometry is in fact already a classical subject [GGMS87]. Just one of such interactions, central for this **book**, is the following set of associations:

$$\text{matroids} \to \text{lattice polytopes} \to \text{toric varieties}.$$

A geometric interpretation of the Tutte polynomial was given in [FS12] via the K-theory of the Grassmannian as follows. Let $Gr(r, n)$ be the Grassmannian of r-dimensional subspaces of an n-dimensional vector space $\mathbb{C}^n$. The torus $T = (\mathbb{C}^*)^n$ acts on $Gr(r, n)$ via its standard action on $\mathbb{C}^n$. A point $L \in Gr(r, n)$ gives rise to a representable of a matroid $M(L)$, by projecting the standard basis of $\mathbb{C}^n$ onto $L^\vee$. The structure sheaf of its torus orbit closure defines a K-class $[\mathcal{O}_{\overline{T \cdot L}}]$ of $Gr(r, n)$ that depends only on the matroid $M(L)$. In general, a matroid M of rank r on a ground set $\{1, \dots, n\}$ defines a K-class $y(M)$ of $Gr(r, n)$. The following theorem of Fink and Speyer relates the K-class $y(M)$ to the Tutte polynomial $T_M(x, y)$ via the diagram

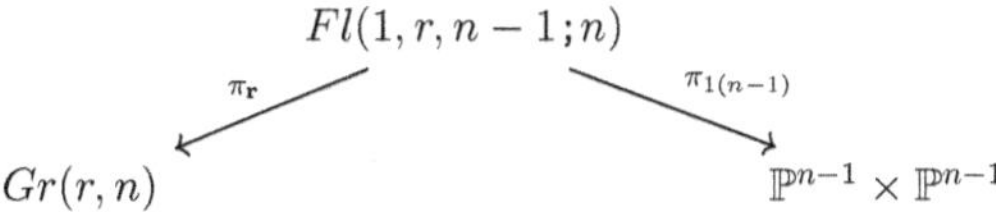

where $Fl(1, r, n - 1; n)$ is a partial flag variety, and π_r and $\pi_{1(n-1)}$ are maps that forget appropriate subspaces in the flag.

Theorem 6.2.12 ([FS12, Theorem 5.1]). *Let* $\mathcal{O}(1)$ *be the line bundle on* $Gr(r; n)$ *of the Plücker embedding* $Gr(r; n) \hookrightarrow \mathbb{P}^{\binom{n}{r} - 1}$. *We have*

$$T_M(\alpha, \beta) = (\pi_{(n-1)1})_* \pi_r^* \left(y(M) \cdot [\mathcal{O}(1)] \right) \in K^0((\mathbb{P}^{n-1})^\vee \times \mathbb{P}^{n-1}) \simeq \mathbb{Z}[\alpha, \beta]/(\alpha^n, \beta^n),$$

where α, β *are* K-*classes of structure sheaves of hyperplanes of* $(\mathbb{P}^{n-1})^\vee, \mathbb{P}^{n-1}$.

The main topic of Part II is a generalization of this relation between the K-theory of Grassmannians and matroids to the relation between K-theory of flag varieties and *flag matroids*. This part is based on joint work with Amanda Cameron, Rodica Dinu, Christopher Eur, Mateusz Michałek [CDMS20, DES20].

Flag matroids are a natural generalization of matroids. They first arose in the work of Gelfand and Serganova [GS87b, GS87a], as a special (type A) case of the so-called *Coxeter matroids*. A flag matroid can be defined as a sequence of matroids on the same ground set that are "compatible" in some sense. Just as a subspace L of a vector space V defines a representable matroid, a flag $L_1 \subseteq \ldots \subseteq L_k$ of subspaces defines a representable flag matroid. An introduction to matroids and flag matroids is given in Chapter 5. In this chapter, we also present a classical result by Gelfand, Goresky, MacPherson and Serganova relating torus orbit closures in Grassmannians with representable matroids, and a generalization to flag varieties and flag matroids.

In Chapter 6, we give an introduction to equivariant K-theory, define the K-class of a matroid, and review the Fink-Speyer construction of the Tutte polynomial. No combinatorial generalization of the Tutte polynomial of a flag matroid is known, but we can use the K-theoretic construction of Fink and Speyer to *define* the *flag-geometric Tutte polynomial* $K\mathcal{T}_{\mathcal{F}}$ of a flag matroid. The combinatorial properties of the flag-geometric Tutte polynomial of a flag matroid are a lot less clear than those of the usual Tutte polynomial of a matroid. In particular, none of the two usual definitions of the Tutte polynomial –the corank-nullity formula and the deletion-contraction relation– seem to generalize to flag matroids.

In Chapter 7, we give an explicit way to compute the flag-geometric Tutte polynomial by summing up Hilbert series of cones. Using techniques from discrete geometry, in particular Brion's formula and cone-flipping, we prove two interesting combinatorial properties of the flag-geometric Tutte polynomial. The first concerns a search for a "corank-nullity formula" for $K\mathcal{T}_{\mathcal{F}}$. A consequence of the corank-nullity formula for a matroid is that one has $T_M(2,2) = 2^n$. We show that the value of $K\mathcal{T}_{\mathcal{F}}(2,2)$ is more intricate.

Theorem 7.3.2. *Let $\mathcal{F}$ be a two-step flag matroid $\mathcal{F} = (M_1, M_2)$ on a ground set E. Let $p\mathcal{B}(\mathcal{F})$ be the set of subsets $S \subseteq E$ such that S is spanning in M_1 and independent in M_2. Then we have*

$$K\mathcal{T}_{\mathcal{F}}(2,2) = 2^{|E|} \cdot |p\mathcal{B}(\mathcal{F})|.$$

The second property concerns a search for analogues of the deletion-contraction recursion. For a two-step flag matroid whose constituent matroids have rank difference 1, we show the following deletion-contraction-like relation.

Theorem 7.3.3. *Let M be a matroid on a ground set E, and let $e \in E$ be neither a loop nor a coloop in M. Then we have*

$$K\mathcal{T}_{(M,M)}(x,y) = K\mathcal{T}_{(M/e,M/e)}(x,y) + K\mathcal{T}_{(M/e,M\backslash e)}(x,y) + K\mathcal{T}_{(M\backslash e,M\backslash e)}(x,y).$$

Notation

Unless otherwise mentioned, we will always be working over the field of complex numbers, and vector spaces will be assumed to be finite-dimensional. In Part II, we will set $[n] := \{1, 2, \ldots, n\}$.

Part I

Tensors

Chapter 1

Introduction to tensors

In this chapter we give a brief introduction to the most fundamental notions related to tensors, in particular: tensor rank, border rank, and their symmetric analogues. We also explain the geometric point of view connecting rank and border rank to secant varieties. We end the chapter with a method for estimating border rank using smoothable schemes, which will play a role in Chapter 3. For a more detailed exposition, we refer the reader to [Lan12].

1.1   Basic definitions

An *order d tensor* (or *d-way tensor*) is simply an element in a tensor product $W := V_1 \otimes \cdots \otimes V_d$, where the V_i are vector spaces. We will denote the dimension of V_i by n_i. If we have chosen a basis $\{e_{i,j} | 1 \leq j \leq n_i\}$ for every V_i, then a basis of W is given by $\{e_{1,j_1} \otimes \cdots \otimes e_{d,j_d} \mid 1 \leq j_i \leq n_i\}$. By writing a tensor $T \in W$ in this basis, we can identify it with an d-dimensional table of scalars. So we can think of tensors as a higher-dimensional generalization of matrices. A fundamental notion in the study of tensors is that of *tensor rank*.

Definition 1.1.1. A tensor $T \in W$ is said to have *rank one* if there are vectors $v_i \in V_i$ such that $T = v_1 \otimes \ldots \otimes v_d$. A *length r rank decomposition* of a tensor $T \in W$ is an expression $T = T_1 + \ldots + T_r$, where the T_i are rank one tensors. The *rank* $\mathrm{rk}(T)$ of a tensor $T \in W$ is the smallest integer r for which there exists a length r rank decomposition.

Since every basis vector $e_{1,j_1} \otimes \cdots \otimes e_{d,j_d}$ has rank one, one immediately sees that every tensor in W has a finite rank at most $\prod_{i=1}^{d} n_i$. In fact, by writing a tensor in our basis, we have written it as a linear combination of rank one tensors $e_{1,j_1} \otimes \cdots \otimes e_{d-1,j_{d-1}} \otimes w$. Hence every tensor has rank at most $\prod_{i=1}^{d-1} n_i$.

The space $\mathrm{Sym}^d(V)$ of symmetric tensors is the subspace of $V^{\otimes d}$ consisting of tensors that are invariant upon permuting the indices. More precisily: we can define the *symmetrization map* $V^{\otimes d} \to V^{\otimes d}$ by linearly extending

$$v_1 \otimes \cdots \otimes v_d \mapsto \frac{1}{d!} \sum_{\sigma \in \mathfrak{S}_d} v_{\sigma(1)} \otimes \cdots \otimes v_{\sigma(d)},$$

and define $\mathrm{Sym}^d(V)$ as the image of the symmetrization map. Note that, as we are working over the algebraically closed field $\mathbb{C}$, the only symmetric rank one tensors are the tensors of the form $v^{\otimes d}$, for $v \in V$.

Definition 1.1.2. A *Waring decomposition* of $T \in \mathrm{Sym}^d(V)$ is an expression $T = v_1^{\otimes d} + \ldots + v_r^{\otimes d}$, where $v_i \in V$. The *Waring rank* $\mathrm{wrk}(T)$ of $T \in \mathrm{Sym}^d(V)$ is the smallest r for which there exists a length r Waring decomposition.

We leave it to the reader to verify that every symmetric tensor has a finite Waring decomposition. Since a Waring decomposition is in particular a rank decomposition, the Waring rank of a symmetric tensor is always greater then or equal to its rank. It was only recently shown [Shi18] that this equality can be strict.

A tensor $T \in V^{\otimes d}$ defines a polynomial map $\phi_T : V^* \to \mathbb{C}$ by putting $\phi_{v_1 \otimes \cdots \otimes v_d}(\beta) = \prod_i \beta(v_i)$ and linearly extending. In coordinates, the tensor $e_{j_1} \otimes \cdots \otimes e_{j_d}$ gives rise to the monomial $x_{j_1} \cdots x_{j_d}$. Clearly, two tensors give rise to the same polynomial if and only if they are the same up to symmetrization. So we can identify the space $\mathrm{Sym}^d(V)$ of symmetric tensors with the space $\mathbb{C}[x_1, \ldots, x_n]_d$ of homogeneous degree d polynomials in n variables. Under this identification, a Waring decomposition of $f \in \mathbb{C}[x_1, \ldots, x_n]_d$ is an expression $f = L_1^d + \cdots + L_r^d$ of f as a sum of d'th powers of linear forms.

Since order 2 tensors are simply matrices, the notion of tensor rank is a generalization of the notion of matrix rank. However, for $d > 2$, the notion of tensor rank is much more ill-behaved than the notion of matrix rank. For instance, the set of $n \times m$ matrices of rank $\leq r$ is a Zariski closed set, defined by the vanishing of $(r+1) \times (r+1)$-minors. On the contrary, the set of tensors of (Waring) rank $\leq r$ is not closed in general. The classical example to illustrate this is the following: let $T \in (\mathbb{C}^2)^{\otimes 3}$ be the following tensor:

$$T = e_1 \otimes e_1 \otimes e_1 + e_1 \otimes e_1 \otimes e_2 + e_1 \otimes e_2 \otimes e_1 + e_2 \otimes e_1 \otimes e_1.$$

One can show that T has rank (and Waring rank) equal to three. However, T can be arbitrarily colosely approximated by (Waring) rank 2 tensors, as follows:

$$T = \lim_{\varepsilon \to 0} \frac{1}{\varepsilon}((\varepsilon - 1)e_1 \otimes e_1 \otimes e_1 + (e_1 + \varepsilon e_2) \otimes (e_1 + \varepsilon e_2) \otimes (e_1 + \varepsilon e_2)).$$

Definition 1.1.3. A *border rank r approximation* of a tensor T is a sequence of tensors T_i of rank $\leq r$ converging to T. The *border rank* $\underline{\mathrm{rk}}(T)$ of T is the smallest r for which a border rank approximation exists. In a similar way, one defines the *border Waring rank* $\underline{\mathrm{wrk}}(T)$ of a symmetric tensor T.

Given a tensor $T \in V_1 \otimes \cdots \otimes V_d$, we can consider for every $i \in \{1, \ldots, d\}$ a linear map

$$T_i : V_i^* \to \bigotimes_{j \neq i} V_j.$$

A border rank approximation of T gives rise to a border rank approximation of the matrix representing T_i. Since for matrices the notions of rank and border rank coincide, this gives a first way of proving lower bounds on the border rank of a tensor: $\underline{\mathrm{rk}}(T) \geq \max_i \mathrm{rk}(T_i)$. If T_i is injective, we say that T is V_i-*concise*; if T is i-concise for every i, we say that T is *concise*. Intuitively, a tensor T not being concise means that we can replace one of the spaces V_i with a strict subspace V_i'. For a concise tensor, we have $\mathrm{rk}(T) \geq \max_i(\dim V_i)$. In case of equality, we say that T has *minimal border rank*.

Definition 1.1.4. If $T \in V_1 \otimes \ldots \otimes V_d$ and $T' \in V_1' \otimes \ldots \otimes V_{d'}'$, then their tensor product $T \otimes T'$ is a $(d + d')$-way tensor in $V_1 \otimes \ldots \otimes V_d \otimes V_1' \otimes \ldots \otimes V_{d'}'$. In the case $d = d'$, the *Kronecker product* $T \boxtimes T'$ is defined as the tensor product of T and T', but viewed as a d-way tensor in $(V_1 \otimes V_1') \otimes \ldots \otimes (V_d \otimes V_d')$. The bracketing is important: the rank one tensors in this space are all tensors of the form $u_1 \otimes \ldots \otimes u_d$, with $u_i \in V_i \otimes V_i'$. So $\mathrm{rk}(T \boxtimes T') \leq \mathrm{rk}(T \otimes T')$, where the inequality can be strict.

Remark 1.1.5. If If $T \in V_1 \otimes \ldots \otimes V_d$ and $T' \in V_1' \otimes \ldots \otimes V_d'$, we can also consider their direct sum $T \oplus T'$, which is a tensor in $(V_1 \oplus V_1') \otimes \ldots \otimes (V_d \oplus V_d')$. It is easy to see that $\mathrm{rk}(T \oplus T') \leq \mathrm{rk}(T) + \mathrm{rk}(T')$. For a long time, it was a conjecture that this was always an equality, but recently, a counterexample has been found by Shitov [Shi19].

1.1.1 Degeneration and restriction of tensors

We work in a space $W = \bigotimes_{i=1}^{n} V_i$. Let $G := GL(V_1) \times \ldots \times GL(V_d)$. There is a natural action of G on V, which extends to an action of the algebra $\mathcal{A} := \mathrm{End}(V_1) \times \ldots \times \mathrm{End}(V_d) \supseteq G$.

Definition 1.1.6. Let $T, T' \in W$.

- We say T' is a *restriction* of T, denoted $T' \leq T$, if $T' \in \mathcal{A} \cdot T$.

- We say T' is a *degeneration* of T, denoted $T' \trianglelefteq T$ if $T' \in \overline{G \cdot T}$.

As every $A \in \mathcal{A}$ can be approximated by elements of G, we get that $T' \leq T \implies T' \trianglelefteq T$. The following proposition follows immediately from the definition.

Proposition 1.1.7. *If T' is a restriction of T, then $\mathrm{rk}(T') \leq \mathrm{rk}(T)$. If T' is a degeneration of T, then $\underline{\mathrm{rk}}(T') \leq \underline{\mathrm{rk}}(T)$.*

In fact, we can *define* rank and border rank in terms of restriction and degeneration, respectively. Let $\langle m \rangle \in (\mathbb{C}^m)^{\otimes d}$ denote the *unit tensor*. Explicitely:

$$\langle m \rangle := \sum_{i=1}^{m} e_i^{\otimes d}.$$

Proposition 1.1.8. *Let $T \in W$, $m \in \mathbb{N}$, and assume that $\dim(V_i) \geq m$ for all i, so that we can view $\langle m \rangle$ as a tensor in W.*

- $T \leq \langle m \rangle \iff \mathrm{rk}(T) \leq m,$

- $T \trianglelefteq \langle m \rangle \iff \underline{\mathrm{rk}}(T) \leq m,$

Proof. The statement about restriction follows immediately from the definitions. For the statement about degeneration, see [Lan17, Section 3.3.1]. $\qquad\square$

Remark 1.1.9. The condition $\dim(V_i) \geq m$ is not essential, as it is always possible to embed each V_i into a higher dimensional vector space V_i'.

Remark 1.1.10. In [BCS97, (15.19)], an alternative definition of degeneration is given, which (in contrast to our definition) works over any field. The equivalence of the two definitions was first proven by Strassen [Str87]; a proof can be found in [BCS97, (20.24)].

Remark 1.1.11. If $T_1 \geq T_1'$ and $T_2 \geq T_2'$, then $T_1 \boxtimes T_2 \geq T_1' \boxtimes T_2'$. Indeed: if $T_1' = (\bigotimes_i f_i) T_1$ and $T_2' = (\bigotimes_i g_i) T_1$ then $T_1' \boxtimes T_2' = (\bigotimes_i (f_i \otimes g_i))(T_1 \boxtimes T_2)$. By taking limits, we also obtain the analoguous statement for degeneration: if $T_1 \trianglerighteq T_1'$ and $T_2 \trianglerighteq T_2'$, then $T_1 \boxtimes T_2 \trianglerighteq T_1' \boxtimes T_2'$. In particular: $\mathrm{rk}(T_1 \boxtimes T_2) \leq \mathrm{rk}(T_1)\,\mathrm{rk}(T_2)$ and $\underline{\mathrm{rk}}(T_1 \boxtimes T_2) \leq \underline{\mathrm{rk}}(T_1)\,\underline{\mathrm{rk}}(T_2)$.

1.2 Geometric perspective

In this section, we will work in the projective spaces $\mathbb{P}(V_1 \otimes \ldots \otimes V_d)$ and $\mathbb{P}(V^{\otimes d})$. In other words, we identify tensors that only differ by a scalar.

Definition 1.2.1. Let $X \subseteq \mathbb{P}^N$ be a variety. The *r-th secant set* of X is the set

$$\sigma_r^\circ(X) := \bigcup_{x_1,\ldots,x_r \in X} \langle x_1, \ldots, x_r \rangle \subseteq \mathbb{P}^N,$$

where $\langle \ldots \rangle$ denotes the linear span. The *r-th secant variety* $\sigma_r(X)$ is the Zariski closure of $\sigma_r^\circ(X)$. For a point $x \in \mathbb{P}^N$, the *X-rank* of x is defined as the smallest r for which $x \in \sigma_r^\circ(X)$. Similarily, the *X-border rank* of x is the smallest r for which $x \in \sigma_r(X)$.

Consider the Segre embedding

$$Seg : \mathbb{P}(V_1) \times \cdots \times \mathbb{P}(V_d) \to \mathbb{P}(V_1 \otimes \ldots \otimes V_d) : ([v_1], \ldots, [v_d]) \mapsto [v_1 \otimes \cdots \otimes v_d].$$

The image $Seg(\mathbb{P}(V_1) \times \cdots \times \mathbb{P}(V_d))$ of Seg is known as the Segre variety. By construction, the Segre variety consists of all rank one tensors in $\mathbb{P}(V_1 \otimes \ldots \otimes V_d)$. Hence the notions of tensor rank and tensor border rank are special cases of the notion of X-rank and X-border rank, with $X = Seg(\mathbb{P}(V_1) \times \cdots \times \mathbb{P}(V_d))$.

Next, we consider the Veronese embedding

$$v_d : \mathbb{P}(V) \to \mathbb{P}(V^{\otimes d}) : [v] \mapsto [v^{\otimes d}].$$

The image $v_d(\mathbb{P}(V))$ of v_d is known as the d-th Veronese variety. By construction, the Veronese variety consists of all Waring rank one tensors in $\mathbb{P}(V^{\otimes d})$. As above, the notions of Waring rank and Waring border rank are special cases of the notion of X-rank and X-border rank, with $X = v_d(\mathbb{P}(V))$. One useful technique for estimating the border rank of a tensor –or more generally, the X-border rank of a point– uses smoothable schemes.

Definition 1.2.2. Let X be a smooth projective variety, and let $R \subseteq X$ be 0-dimensional subscheme of degree r. We say that R is *smoothable in X*, if R can be written as a flat limit of smooth schemes inside the Hilbert scheme of 0-dimensional degree r subschemes of X.

In the rest of this section, R and R_t will always stand for zero-dimensional schemes of degree r. Since smooth 0-dimensional schemes of degree r are simply r-tuples of points, R being smoothable means that there are families $x_i(t)$ of points on X such that

$$R = \lim_{t \to 0} \{x_1(t), \ldots, x_r(t)\}.$$

The following lemma shows that being smoothable is a property of the scheme, not of the embedding.

Lemma 1.2.3 ([BB14, Proposition 2.1], see also [BJ17, Theorem 3.16]). *Let R be a zero-dimensional degree r scheme, and let $R \hookrightarrow X$ and $R \hookrightarrow Y$ be two embeddings of R into a smooth variety. If R is smoothable in X, then R is smoothable in Y.*

Now suppose we have a smoothable subscheme $R \subseteq X \subseteq \mathbb{P}^N$. Then $R = \lim_{t \to 0} R_t$ for some flat family R_t of smooth subschemes of X. More precesily, there is a variety T and a closed subscheme Y of $X \times T$ that is flat over T, with the fiber of $Y \to T$ over $t \in T$ being equal to R_t and the fiber over $0 \in T$ equal to R. Outside of a closed subset of T, the linear span $\langle R_t \rangle$ is of constant dimension q. So we can consider the limit $\lim_{t \to 0} \langle R_t \rangle$ in the Grassmannian $\mathbb{G}(q, \mathbb{P}^N)$. If a point $x \in \mathbb{P}^N$ is contained in the limit of the linear spans $\lim_{t \to 0} \langle R_t \rangle$, then $\underline{\mathrm{rk}}_X(x) \leq r$. We can compare $\lim_{t \to 0} \langle R_t \rangle$ with the scheme-theoretic linear span $\langle R \rangle = \langle \lim_{t \to 0} R_t \rangle$.

Proposition 1.2.4 (See [BJ17, Proposition 5.29]). *Let R_t be a flat family of 0-dimensional degree r subschemes of $\mathbb{P}^N$. Then*

$$\langle \lim_{t\to 0} R_t \rangle \subseteq \lim_{t\to 0} \langle R_t \rangle.$$

Here, the limit on the left hand side is taken in the Hilbert scheme of 0-dimensional degree r subschemes of $\mathbb{P}^N$, and the limit on the right hand side is taken in the appropriate Grassmannian.

Proof. Let $S = \mathbb{C}[x_0, \ldots, x_N]$ and write $R_t = \operatorname{Proj} S/I_t$ and $\lim_{t\to 0} R_t = \operatorname{Proj} S/I$, where I_t and I are saturated ideals. By the usual construction of the Hilbert scheme, there exists a constant m such that for every 0-dimensional degree r subscheme $\operatorname{Proj} S/J$ of $\mathbb{P}^N$, the degree m part $J(m)$ of J has the same dimension $c := \binom{N+m}{m} - r$. Then the ideal I satisfies $I(m) = \lim_{t\to 0} I_t(m)$, where the limit is taken in the Grassmannian $Gr\left(c, \binom{N+m}{m}\right)$. The linear span of a subscheme $\operatorname{Proj} S/J \subseteq \mathbb{P}^N$ is cut out by $J(1)$. Hence we need to show that $\lim_{t\to 0} I_t(1) \subseteq I(1)$. Let $f \in \lim_{t\to 0} I_t(1)$. Then there is a family $f_t \in I_t(1)$ such that $f = \lim_{t\to 0} f_t$. For every $g \in S(m-1)$, it holds that $gf = \lim_{t\to 0} gf_t \in \lim_{t\to 0} I_t(m) = I(m)$. Since I is saturated, it follows that $f \in I(1)$. $\qquad\square$

Corollary 1.2.5. *If there exists a smoothable 0-dimensional subscheme $R \subset X$ of degree r, such that $x \in \langle R \rangle$, then $\underline{\mathrm{rk}}_X(x) \leq r$.*

There are several results in the literature that allow us to prove that a scheme is smoothable, without having to explicitly write it as a limit of smooth schemes. Combining these with Corollary 1.2.5 can be very useful for proving estimates on the border rank of a given $x \in X$. We will use this method in Section 3.3.

Example 1.2.6. The inequality in Proposition 1.2.4 can be strict. We give an example in affine space $\mathbb{A}^2 \subset \mathbb{P}^2$. Consider $x_1(t) = (0,0)$, $x_2(t) = (t,0)$, and $x_3(t) = (t^2, t^4)$ in $\mathbb{A}^2$. For $t \neq 0$, it holds that $\langle R_t \rangle = \mathbb{A}^2$. But one can compute

$$\lim_{t\to 0}\left(\langle x, y \rangle \cap \langle x-t, y \rangle \cap \langle x-t^2, y-t^4 \rangle\right) = \langle x^3, y \rangle,$$

so we find

$$\langle \lim_{t\to 0} R_t \rangle = V(y) \subsetneq \mathbb{A}^2 = \lim_{t\to 0} \langle R_t \rangle.$$

Remark 1.2.7. The *smoothable X-rank* of a point $x \in \mathbb{P}^n$ is defined as the smallest r such that there is a smoothable 0-dimensional subscheme $R \subset X$ of degree r with $x \in \langle R \rangle$. Corollary 1.2.5 says that smoothable rank is greater than or equal to border rank, and as Example 1.2.6 suggests, equality can be strict. There is also the notion of *cactus rank*, where we consider arbitrary 0-dimensional schemes instead of smoothable ones. A lot of research has been done comparing these various notions of rank, see for example [RS11, BB14, BBM14, BB15].

Chapter 2

Matrix product states

In this chapter, we apply methods from algebraic geometry to study uniform matrix product states. Uniform matrix product states are certain tensors which arise in quantum information theory: they model quantum systems consisting of a number of sites placed on a ring.

In Section 2.1, we give a brief introduction to tensor networks. The main goal is to motivate the definition of uniform matrix product states and to place it in a more general context. The reader willing to accept (2.2.1) at face value may postpone reading this section. However, the language developed will be used again in Section 2.5. In Section 2.2 we collect basic definitions, results and notations regarding uniform matrix product states $\mathrm{uMPS}(D, d, N)$. We discuss the closedness in trivial cases, dimension, and local symmetries.

Section 2.3 is an interlude about the injectivity radius and generic injectivity radius. The results in this section can be stated in elementary terms, but have various applications to the theory of matrix product states. In particular, we recall the quantum Wielandt theorem, and give a short nonconstructive proof.

The next Section 2.4 contains our main results on uniform matrix product states. It consists of four parts. In Section 2.4.1 we give a complete classification when $\mathrm{uMPS}(2, d, N)$ is closed. Then we discuss closedness in other cases, and we prove connectedness of the uMPS. In Section 2.4.2 we explore for which parameters the set $\mathrm{uMPS}(D, d, N)$ fills the ambient space $\mathrm{Cyc}^N(\mathbb{C}^d)$. In Section 2.4.3 we recall another parametrization of matrix product states. Using this *trace parametrization* and Macaulay2 [GS] we obtain defining equations for $\mathrm{uMPS}(2, 2, N)$ for small values of N. Section 2.4.4 is devoted to new results related to identifiability and the so-called fundamental theorem of matrix product states. Finally, in Section 2.4.5 we give a full description of $\mathrm{uMPS}(2, 2, 4)$ as a constructible subset of the ambient space $\mathrm{Cyc}^4(\mathbb{C}^2)$.

The final Section 2.5 is devoted to a generalization of the quantum Wielandt theorem. This theorem has applications to a class of tensor network states called *projected entangled pair states* (PEPS), which generalize matrix product states.

The results in Sections 2.1 and 2.3 were known before, though the proof of

Theorem 2.3.7 given here is new. Sections 2.2 and 2.4 are based on the paper [CMS19], and Section 2.5 is based on [MSV19].

2.1 Tensor networks

This section consists of a brief introduction to tensor networks. The study of tensor networks is motivated by quantum information theory, where they represent quantum systems of sites placed on a graph.

2.1.1 Contraction, and pictorial representation of tensors

Contracting tensors is the natural generalization of matrix multiplication. If $W = V_1 \otimes \cdots \otimes V_n$ and $W' = V_1' \otimes \cdots \otimes V_m'$ are tensor spaces satisfying $V_1' = V_1^*$, then there is a natural bilinear map $W \times W' \to (V_2 \otimes \cdots \otimes V_n) \otimes (V_2' \otimes \cdots \otimes V_m')$, defined on rank one tensors by $(v_1 \otimes \cdots \otimes v_n, v_1' \otimes \cdots \otimes v_m') \mapsto \langle v_1, v_1' \rangle (v_2 \otimes \cdots \otimes v_n \otimes v_2' \otimes \cdots \otimes v_m')$. We can also view this in coordinates: choose bases for the V_i and V_i', such that the chosen basis of V_1 is dual to the chosen basis of V_1'. If $T \in W$ has coordinates $T_{i_1 \ldots i_n}$, and $T' \in W'$ has coordinates $T'_{j_1 \ldots j_m}$, then their contraction C has coordinates $C_{i_2 \ldots i_n, j_2 \ldots j_m} = \sum_k T_{k, i_2 \ldots i_n} T'_{k, j_2 \ldots j_m}$, in other words we "sum out the first index".

A usefel pictorial way of thinking about contraction of tensors is to picture a d-way tensor as a box with d outgoing edges, labelled by the vector spaces V_i. The contraction of two tensors is then denoted by joining the edges together. For instance, the multiplication of two matrices $A \in U^* \otimes V$ and $B \in V^* \otimes W$ would be written as

$$U^* \boxed{A} \;\; V \;\; V^* \boxed{B} \;\; W \; .$$

2.1.2 Tensor network states

Definition 2.1.1. A *directed multigraph with outgoing edges* (henceforth: graph) is a quadruple $(\mathcal{V}, E, s, t)$, where

- $\mathcal{V}$ and E are finite sets, called *vertices* and *edges*,

- $s : E \to \mathcal{V} \sqcup \{\infty\}$ assigns to every edge its source,

- $t : E \to \mathcal{V} \sqcup \{\infty\}$ assigns to every edge its target,

such that there is no edge e with $s(e) = t(e) = \infty$.

If $s(e) = \infty$ or $t(e) = \infty$, we call e an *outgoing edge*; if $s(e) \in \mathcal{V}$ and $t(e) \in \mathcal{V}$, we call e an *inner edge*.

A tensor network is determined by the following data:

- a graph $\Gamma = (\mathcal{V}, E, s, t)$, and

- for every edge $e \in E$ a finite-dimensional vector space V_e.

To every vertex $p \in \mathcal{V}$, we can associate the vector space $V_p := (\bigotimes_{e:s(e)=p} V_e) \otimes (\bigotimes_{e:t(e)=p} V_e^*)$. Furthermore, define $W_\Gamma := (\bigotimes_{e:s(e)=\infty} V_e^*) \otimes (\bigotimes_{e:t(e)=\infty} V_e)$ [1]. Consider the map

$$\bigoplus_{p \in \mathcal{V}} V_p \to W_\Gamma \tag{2.1.1}$$

given by contracting a collection of tensors $T_p \in V_p$ along the edges of Γ. We say that a vector $w \in W_\Gamma$ is a *tensor network state* for the pair $(\Gamma, \{V_e\})$, if w is in the image of this map.

Remark 2.1.2. In practice, one often fixes an identification of every vector space V_e with its dual. This allows us to work with undirected graphs instead of directed graphs. In particular, in quantum information theory (see Remark 2.1.4), one considers Hilbert spaces, which come equipped with an inner product that identifies the space with its dual.

Example 2.1.3 (Matrix Product States). Fix a natural number $N \in \mathbb{N}$, and tuples $d = (d_1, \ldots, d_N) \in \mathbb{N}^N$, $D = (D_1, \ldots, D_{N-1}) \in \mathbb{N}^{N-1}$. Let Γ be the following graph:

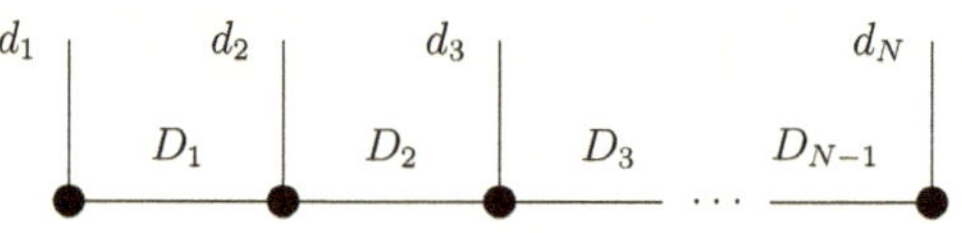

where we labeled every edge e with the dimension of the corresponding space V_e. We have $V_{p_1} = \mathbb{C}^{d_1} \otimes \mathbb{C}^{D_1}$, $V_{p_N} = \mathbb{C}^{D_{N-1}} \otimes \mathbb{C}^{d_N}$, and $V_{p_i} = \mathbb{C}^{D_{i-1}} \otimes \mathbb{C}^{d_i} \otimes \mathbb{C}^{D_i}$ for $1 < i < N$. We can identify a tensor $T_i \in V_{p_i}$ with a d_i-tuple of $D_{i-1} \times D_i$-matrices $(M_0^i, \ldots, M_{d_i-1}^i)$, where we put $D_0 = D_N = 1$. Then (2.1.1) becomes:

$$\bigoplus_i (\mathbb{C}^{D_{i-1} \times D_i})^{d_i} \to \bigotimes_i \mathbb{C}^{d_i} \tag{2.1.2}$$

$$(\{M_j^1\}_{0 \le j < d_1}, \ldots, \{M_j^N\}_{0 \le j < d_N}) \mapsto \sum_{j_1, \ldots, j_N} (M_{j_1}^1 \cdots M_{j_N}^N) e_{j_1} \otimes \cdots \otimes e_{j_N}. \tag{2.1.3}$$

The number of parameters needed to represent a matrix product state is equal to $\sum_i d_i D_{i-1} D_i$. For large values of N, this is much smaller then the dimension d^N of our tensor space.

---

[1] Note that W_Γ depends not just on Γ, but also on the V_e. We suppress this to make the notation less cumbersome.

Remark 2.1.4. The motivation for considering tensor networks comes from quantum information theory (QIT). What follows is a brief explanation of this motivation. We refer the reader to [CGRPG19, Section 1] for more details.

In QIT, a state of a simple system is represented by a vector[2] v in a finite-dimensional Hilbert space $\mathbb{C}^d$. A state of a composite system, consisting of N sites, is a tensor in the tensor product $V = \mathbb{C}^{d_1} \otimes \cdots \otimes \mathbb{C}^{d_N}$, where $\mathbb{C}^{d_i}$ is the Hilbert space corresponding to the i-th site. A rank one tensor $v_1 \otimes \cdots \otimes v_n \in V$ describes a state in which the i-th site is in state v_i. However, the space V also contains tensors of higher rank. These arise from the quantum-mechanical phenomenon of *entanglement*. As an example, consider a tensor $\frac{1}{\sqrt{2}}(v_1 \otimes v_1 + v_2 \otimes v_2) \in \mathbb{C}^d \otimes \mathbb{C}^d$. This represents a state in which with probability $\frac{1}{2}$, both sites are in state v_1, and with probability $\frac{1}{2}$, both sites are in state v_2.

If the number of particles is large, the tensor space V will be of extremely high dimension. Therefore, it is useful to restrict to a subset of states which are physically meaningful. Tensor network states provide one way of doing this. Assume that at every vertex v of Γ there is exactly one outgoing edge. Then we can identify the vertices of Γ with the sites of a physical system; the tensor space W_Γ is then the state space of this system. We can think of tensor network states as states with limited entanglement, where the dimension of the vector space V_e associated to the inner edge e indicated how much entanglement we allow between its endpoints. Note that if for every inner edge, V_e is one dimensional, then the tensor network states are precisely the rank one tensors in W_Γ, i.e. the states with no entanglement.

Remark 2.1.5. Tensor networks also play a role in computational mathematics, where they arise as certain *tensor formats*, used to compress high-dimensional data. In this context, matrix product states are known under the name *tensor train format*, a term introduced by Oseledets [Ose11]. We refer the reader to [Hac12] for a survey.

The set $TNS(\Gamma)$ of tensor network states associated to a pair $(\Gamma, \{V_e\})$ is by construction the image if a polynomial map. Hence its Zariski closure $\overline{TNS(\Gamma)}$ equals its Euclidean closure and is an algebraic variety. In [LQY12], $TNS(\Gamma)$ and $\overline{TNS(\Gamma)}$ were studied from a geometric perspective. In particular, it was shown that if Γ is a tree, then $TNS(\Gamma)$ and $\overline{TNS(\Gamma)}$ coincide, but for arbitrary Γ they can be different.

2.1.3 Uniform tensor network states

Intuitively, uniform tensor networks are tensor networks where we identify all vertices of Γ, and insist on placing the same tensor at every vertex. They are used to represent states with a translational symmetry. In this section, we make

---

[2] Typically, it is assumed that v has norm one, but we will not make this restriction.

this more precise, and we introduce *uniform matrix product states* (uMPS) – the central object of study in this chapter– as a special case of uniform matrix product states. We also introduce the higher-dimensional analogue of uMPS, called *projected entangled pair states* (PEPS). These will play a central role in Section 2.5.

A uniform tensor network is determined by the following data:

- a graph $\Gamma = (\mathcal{V}, E, s, t)$,

- a group G acting on Γ (i.e. G acts on $\mathcal{V}$ and E, such that $s(g \cdot e) = g \cdot s(e)$ and $t(g \cdot e) = g \cdot t(e)$ for all $g \in G$ and $e \in E$. We put $g \cdot \infty = \infty$). We assume that G acts strictly transitively on the vertices $\mathcal{V}$ of G,

- for every G-orbit $\bar{e}$ in E a finite-dimensional vector space $V_{\bar{e}}$.

Let $v \in \mathcal{V}$ be any vertex, and define $V := (\bigotimes_{e:s(e)=v} V_{\bar{e}}) \otimes (\bigotimes_{e:t(e)=v} V_{\bar{e}}^*)$. Since G acts stricly transitively on $\mathcal{V}$, the space V is independent of the chosen vertex V. Define $W_\Gamma := (\bigotimes_{e:s(e)=\infty} V_e^*) \otimes (\bigotimes_{e:t(e)=\infty} V_e)$ as before.

Consider the map

$$V \twoheadrightarrow \bigotimes_{v \in \mathcal{V}} V \twoheadrightarrow W_\Gamma$$

given by taking a tensor $T \in V$, placing it at every vertex of Γ, and contracting along the edges of Γ. We say that a vector $w \in W_\Gamma$ is a *uniform tensor network state* for the triple $(\Gamma, G, \{V_{\bar{e}}\})$, if w is in the image of this map.

Definition 2.1.6 (Uniform matrix product states). Fix natural numbers D (bond dimension), d (physical dimension) and N (number of sites), and let Γ be the following graph, with N vertices:

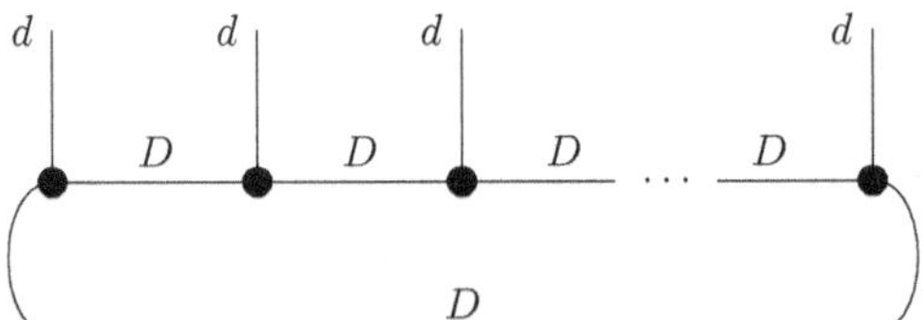

The cyclic group $\mathbb{Z}/N\mathbb{Z}$ acts on Γ in an obvious way, which allows us to identify the vertices of Γ. The uniform tensor network states of $(\Gamma, \mathbb{Z}/N\mathbb{Z}, (\mathbb{C}^D, \mathbb{C}^d))$ are called *uniform matrix product states*. The defining map (henceforth called "uMPS map") takes the form $(\mathbb{C}^D)^{\otimes 2} \otimes \mathbb{C}^d \to (\mathbb{C}^d)^{\otimes N}$. By identifying tensors in $(\mathbb{C}^D)^{\otimes 2} \otimes \mathbb{C}^d$ with d-tuples of $D \times D$ matrices, the uMPS map can be viewed as a polynomial map

$$T_N : (\mathbb{C}^{D \times D})^d \to (\mathbb{C}^d)^{\otimes N}$$

$$(M_0, \ldots, M_{d-1}) \mapsto \sum_{0 \leq i_1, \ldots, i_N \leq d-1} \mathrm{tr}(M_{i_1} \cdots M_{i_N}) e_{i_1} \otimes \cdots \otimes e_{i_N}.$$

21

Uniform matrix product states represent cyclically symmetric states of N identical sites placed on a ring.

Definition 2.1.7 (PEPS). Fix two natural numbers D and d, and an n-tuple $(N_1, \ldots, N_n)$. Our graph Γ is an n-dimensional grid on an n-dimensional torus, of size $N_1 \times \cdots \times N_n$, with one outgoing edge at every vertex. The picture below shows Γ in the case $n = 2, N_1 = 3, N_2 = 5$:

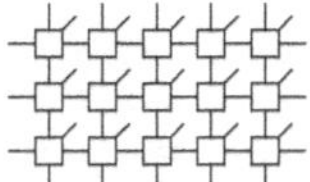

where the edges at top and bottom, respectively left and right, are connected. The group G equals $\mathbb{Z}/N_1\mathbb{Z} \times \cdots \times \mathbb{Z}/N_n\mathbb{Z}$. At every inner edge we put the space $\mathbb{C}^D$, at every outgoing edge we put the space $\mathbb{C}^d$. The uniform tensor network states of $(\Gamma, G, (\mathbb{C}^D, \mathbb{C}^d))$ are called *projected entangled pair states*. In the case $n = 1$, we recover the previous definition.

2.2 Uniform matrix product states: basic properties

In this section we will state some basic properties of uniform matrix product states. We fix 3 parameters $D, d, N \in \mathbb{N} \setminus \{0\}$. Recall from the previous section that a tensor $T \in (\mathbb{C}^d)^{\otimes N}$ is called a *uniform matrix product state* for the triple (D, d, N) if there is a collection of d matrices $M_0, \ldots, M_{d-1}$ in $\mathbb{C}^{D \times D}$ such that

$$T = T_N(M_0, \ldots, M_{d-1}) := \sum_{0 \leq i_1, \ldots, i_N \leq d-1} \mathrm{tr}(M_{i_1} \cdots M_{i_N}) e_{i_1} \otimes \cdots \otimes e_{i_N}. \quad (2.2.1)$$

We will sometimes write $T_{D,d,N}$ instead of T_N if the parameters are not clear from the context. The set of all uniform D-matrix product states in $(\mathbb{C}^d)^{\otimes N}$ is denoted by $\mathrm{uMPS}(D, d, N)$. In other words, $\mathrm{uMPS}(D, d, N)$ is the image of the polynomial map

$$T_N : (\mathbb{C}^{D \times D})^d \to (\mathbb{C}^d)^{\otimes N}.$$

From now on we will leave the triple (D, d, N) implicit and simply use the terminology *uniform matrix product state*. We point out that every cyclically symmetric tensor in $(\mathbb{C}^d)\otimes N$ will be a matrix product state for D large enough (this will follow from Corollary 2.4.16).

Remark 2.2.1. In Section 2.1, we motivated matrix product states with tensor networks and quantum information theory. A different motivation is the following: note that for $D = 1$, the uMPS map is (after projectivization) precisely the

Veronese embedding. Thus we can think of the uMPS map as a kind of *non-commutative Veronese map*, where we replaced the product of scalars with the product of $D \times D$ matrices.

The set $\mathrm{uMPS}(D, d, N)$ is a cone, i.e. if $T \in \mathrm{uMPS}(D, d, N)$, then also $\lambda T \in \mathrm{uMPS}(D, d, N)$ for every $\lambda \in \mathbb{C}$. This is no longer true if we replace $\mathbb{C}$ by a field that is not algebraically closed, e.g. for $\mathbb{R}$ it is only guaranteed if N is odd or $\lambda \geq 0$. Since $\mathrm{tr}(M_{i_1} \cdots M_{i_N})$ does not change if we cyclically permute the matrices in the product, it follows that $\mathrm{uMPS}(D, d, N) \subseteq \mathrm{Cyc}^N(\mathbb{C}^d)$, where $\mathrm{Cyc}^N(\mathbb{C}^d) \subseteq (\mathbb{C}^d)^{\otimes N}$ is the subspace of cyclically symmetric tensors.

As $\mathrm{uMPS}(D, d, N)$ is the image of a polynomial map, it is a constructible set (i.e. a finite union of locally Zariski closed sets) by Chevalley's theorem [DG71, IV, 1.8.4.]. Its Euclidean closure $\overline{\mathrm{uMPS}(D, d, N)}$ agrees with its Zariski closure and is an algebraic variety. In Section 2.4.3 we will give defining equations for small parameter values, and in Section 2.4.5 we give a complete description of the smallest nontrivial case $\mathrm{uMPS}(2, 2, 4)$. A natural question to ask is the following.

Question 2.2.2. *For which parameters D, d, N is $\mathrm{uMPS}(D, d, N)$ a closed set?*

Analogous questions have been investigated from the point of view of complex and real tensors of bounded rank. In that case, most often, the locus is not closed, leading to the central notion of border rank [Lan12, Lan17, DSL08, QML19, SS17, BB14]. Question 2.2.2 will be the main subject of Section 2.4.1. Below we collect some easy results regarding closedness of $\mathrm{uMPS}(D, d, N)$. The next lemma follows immediately from the definitions.

Lemma 2.2.3. *If $D \leq D'$, then $\mathrm{uMPS}(D, d, N) \subseteq \mathrm{uMPS}(D', d, N)$.*

Corollary 2.2.4. *If $\mathrm{uMPS}(D, d, N) = \mathrm{Cyc}^N(\mathbb{C}^d)$, then for any $D' \geq D$, $\mathrm{uMPS}(D', d, N) = \mathrm{Cyc}^N(\mathbb{C}^d)$. In particular, in such a case $\mathrm{uMPS}(D', d, N)$ is closed for all $D' \geq D$.*

Lemma 2.2.5. *For $d \leq d'$ we have an inclusion $(\mathbb{C}^d)^{\otimes N} \subset (\mathbb{C}^{d'})^{\otimes N}$. The following equality holds:*

$$\mathrm{uMPS}(D, d, N) = \mathrm{uMPS}(D, d', N) \cap (\mathbb{C}^d)^{\otimes N}.$$

Proof. The inclusion $\subseteq$ follows from the equality

$$T_N(M_0, \ldots, M_{d-1}) = T_N(M_0, \ldots, M_{d-1}, 0, \ldots, 0).$$

For the other inclusion we note that the projection $(\mathbb{C}^{d'})^{\otimes N} \to (\mathbb{C}^d)^{\otimes N}$ maps $\mathrm{uMPS}(D, d', N)$ to $\mathrm{uMPS}(D, d, N)$. $\square$

Corollary 2.2.6. *If $\mathrm{uMPS}(D, d, N)$ is not closed, then for all $d' \geq d$, the set $\mathrm{uMPS}(D, d', N)$ is not closed.*

Proposition 2.2.7. *If $D = 1$ or $d = 1$ or $N \leq 2$, then* $\mathrm{uMPS}(D, d, N)$ *is closed.*

Proof. If $d = 1$ or $N = 1$, then by definition $\mathrm{uMPS}(D, d, N) = (\mathbb{C}^d)^{\otimes N}$.
For $D = 1$: $\mathrm{uMPS}(1, d, N)$ is equal to the image of the Veronese embedding
$\mathbb{C}^d \to \mathrm{Sym}^N(\mathbb{C}^d) \subset (\mathbb{C}^d)^{\otimes N}$, which is known to be closed.
For $N = 2$: $\mathrm{uMPS}(D, d, 2) \subseteq (\mathbb{C}^d)^{\otimes 2} = \mathbb{C}^{d \times d}$ consists of all symmetric $d \times d$-
matrices of rank at most D^2, which is a closed set. $\qquad\square$

If $T \in \mathrm{uMPS}(D, d, N)$, there can be many different choices of the matrices
$M_0, \ldots, M_{d-1}$ exhibiting T as $T_N(M_0, \ldots, M_{d-1})$. In particular, we have the
following.

Remark 2.2.8. Observe that for $P \in GL(D, \mathbb{C})$, it holds that

$$T_N(M_0, \ldots, M_{d-1}) = T_N(P^{-1} M_0 P, \ldots, P^{-1} M_{d-1} P).$$

In particular, for $T \in \mathrm{uMPS}(D, d, N)$, we can write $T = T_N(M_0, \ldots, M_{d-1})$ where
M_0 is in Jordan normal form.

We expect that the generic fiber of the uMPS-map $(\mathbb{C}^{D \times D})^d \to \mathrm{Cyc}^N(\mathbb{C}^d)$
consists of $D!\,N$ simultaneous conjugacy classes: the $D!$ comes from permuting
the rows and columns of the matrices, and the N from multiplying each matrix
with an N-th root of unity. In Section 2.4.4, we will show this is true for large
N. It is a corollary of the *fundamental theorem of uniform matrix product states*
[CPGSV17, MGRPG$^+$18]. We now discuss the expected dimension of the uMPS.

Proposition 2.2.9. *The dimension of the variety* $\overline{\mathrm{uMPS}(D, d, N)}$ *is at most*
$\min\{(d - 1)D^2 + 1, \dim(\mathrm{Cyc}^N(\mathbb{C}^d))\}$.

Proof. If $T \in \mathrm{uMPS}(D, d, N)$, we can write $T = T_N(M_0, \ldots, M_{d-1})$. By Remark 2.2.8 we can, for M_0 and M_1 generic, assume that M_0 is diagonal and that
the $D - 1$ nondiagonal entries on the top row of M_1 are all equal to 1. Since the
dimension of the image of a polynomial map does not change if we restrict to a
dense subset, we are done. $\qquad\square$

Remark 2.2.10. It is not hard to show that the dimension of the ambient space
$\mathrm{Cyc}^N(\mathbb{C}^d)$ is equal to

$$\dim \mathrm{Cyc}^N(\mathbb{C}^d) = \frac{1}{N} \sum_{\ell \mid N} \varphi(\ell) d^{N/\ell}$$

where φ is the Euler totient function. If N is prime, this simplifies to

$$\dim \mathrm{Cyc}^N(\mathbb{C}^d) = \frac{d^N + (N - 1)d}{N}.$$

Definition 2.2.11. If the equality

$$\dim \mathrm{uMPS}(D, d, N) = \min\{(d-1)D^2 + 1, \dim(\mathrm{Cyc}^N(\mathbb{C}^d))\}$$

holds, we say that $\mathrm{uMPS}(D, d, N)$ has *expected dimension*.

It is natural to ask for which parameters $\mathrm{uMPS}(D, d, N)$ has expected dimension. This question is very similar to the problem of determining for which parameters the set of tensors of bounded tensor rank has expected dimension, which has received considerable attention in the literature. The most famous result along these lines is the celebrated Alexander-Hirschowitz Theorem [AH95], which gives a complete answer in the case of symmetric tensor rank. For general tensors the question remains open — for important partial results we refer to [CGG11, Lan12, OS16, AB12].

Remark 2.2.12. The dimension of $\overline{\mathrm{uMPS}(D, d, N)}$ is easy to compute for small values of (D, d, N) using the Jacobian criterion. In particular, we verified that for all cases in Table 1 in the introduction, $\mathrm{uMPS}(D, d, N)$ has expected dimension.

Conjecture 2.2.13. *For every choice of (D, d, N), $\mathrm{uMPS}(D, d, N)$ has expected dimension.*

We can obtain more cases for which Conjecture 2.2.13 holds from our results in Section 2.4. More precisely: $\mathrm{uMPS}(D, d, N)$ clearly has expected dimension if its closure fills the space $\mathrm{Cyc}^N(\mathbb{C}^d)$. In Corollary 2.4.16, this is shown to hold for large D. Additionally, $\mathrm{uMPS}(D, d, N)$ has expected dimension if the general fiber of the map T_N has dimension $D^2 - 1$. This holds for large N by the fundamental theorem (see Corollary 2.4.24).

Remark 2.2.14. The set $\mathrm{uMPS}(D, d, N)$ is invariant under *local symmetries*. Explicitely, note that the action of GL_d on $\mathbb{C}^d$ induces an action of GL_d on $(\mathbb{C}^d)^{\otimes N}$ by $A \cdot (v_1 \otimes \ldots \otimes v_N) = (A \cdot v_1) \otimes \ldots \otimes (A \cdot v_N)$. This action restricts to the space $\mathrm{Cyc}^N(\mathbb{C}^d)$. It is compatible with the uMPS map in the sense that for a matrix $A = (a_{i,j})_{ij} \in GL_d$ and $(M_0, \ldots, M_{d-1}) \in (\mathbb{C}^{D \times D})^d$, it holds that

$$T_N\Big(\sum_j a_{0,j} M_j, \ldots, \sum_j a_{d-1,j} M_j\Big) = A \cdot T_N(M_0, \ldots, M_{d-1}).$$

This implies that $\mathrm{uMPS}(D, d, N)$ is invariant under the action defined above.

We conclude this section by introducing some notation. Let us denote $[d] = \{0, \ldots, d-1\}$. Then for $(i_1, \ldots, i_N) \in [d]^N$, we define

$$e_{i_1 i_2 \cdots i_N} := \sum_{(j_1, \ldots, j_N)} e_{j_1} \otimes \cdots \otimes e_{j_N} \in \mathrm{Cyc}^N(\mathbb{C}^d) \subseteq (\mathbb{C}^d)^{\otimes N},$$

where the sum is over all disctinct cyclic permutations of $(i_1, \ldots, i_N)$. For example: $e_{111} = e_1 \otimes e_1 \otimes e_1$, $e_{110} = e_1 \otimes e_1 \otimes e_0 + e_1 \otimes e_0 \otimes e_1 + e_0 \otimes e_1 \otimes e_1$, and $e_{1010} = e_1 \otimes e_0 \otimes e_1 \otimes e_0 + e_0 \otimes e_1 \otimes e_0 \otimes e_1$.

Definition 2.2.15. The W-*state* is defined as

$$W_N := e_{0\cdots01} \in \mathrm{Cyc}^N(\mathbb{C}^2).$$

The W-state plays, for example, an important role in Quantum Information Theory [CDV00]. It will also be essential in our discussion about closedness in Section 2.4.1.

2.3 Injectivity radius and generic injectivity radius

Let $D, d \geq 2$ be natural numbers. The injectivity radius $C_{D,d}$ and generic injectivity radius $GC_{D,d}$ are two constants (depending on D, d) that appear in various places in the theory of matrix product states. However, they also arise naturally when considering elementary questions about spaces spanned by products of matrices. In this section, we will use this elementary approach to define $C_{D,d}$ and $GC_{D,d}$ and review what is known about them. The connections to matrix product states will appear in the later sections of this chapter.

In this section, we will be working in a space $V \cong (\mathbb{C}^D)^{\otimes 2} \otimes \mathbb{C}^d$. Tensors in V will be identified with d-tuples of $D \times D$ matrices. For $A \in V$, its slices will be denoted by $A_1, \ldots, A_d \in (\mathbb{C}^D)^{\otimes 2}$; the linear span of these slices will be denoted by $\langle A \rangle \subseteq (\mathbb{C}^D)^{\otimes 2}$. Given $A \in V$ and $N \in \mathbb{N}$, we define the space $A^N \subseteq (\mathbb{C}^D)^{\otimes 2}$ to be the linear span of N-fold products of matrices in $\langle A \rangle$. Equivalently,

$$A^N := \langle \{A_{i_1} \cdot A_{i_2} \cdots A_{i_N} | (i_1, \ldots, i_N) \in [d]^N \} \rangle$$

We say that $A \in V$ is *spanning*, if there is an N such that $A^N = (\mathbb{C}^D)^{\otimes 2}$. The smallest such N is called the *spanning index* of A, which we will denote by $i(A)$. If A is not spanning, we say that $i(A) = \infty$. Note that $i(A)$ only depends on $\langle A \rangle$. Because of this, we will from now on assume that $d \leq D^2$. Note that it is not always true that $A^{N+1} \supseteq A^N$ (consider for instance the case where A consists of strictly upper triangular matrices). However, the following is still true.

Lemma 2.3.1. *If $A^N = (\mathbb{C}^D)^{\otimes 2}$, then also $A^{N+1} = (\mathbb{C}^D)^{\otimes 2}$. In other words: for every $N \geq i(A)$, it holds that $A^N = (\mathbb{C}^D)^{\otimes 2}$.*

Proof. Let $B \in (\mathbb{C}^D)^{\otimes 2}$. Then by assumption B can be written as a linear combination $B = \sum_k \lambda_k A_{i_{(k,1)}} \cdots A_{i_{(k,N)}}$. But every product $A_{i_{(k,1)}} \cdots A_{i_{(k,N-1)}}$ can, again by assumption, be rewritten as a linear combination of the form $\sum_\ell \mu_{k,\ell} A_{j_{(k,\ell,1)}} \cdots A_{j_{(k,\ell,N)}}$. Hence

$$B = \sum_{k,\ell} \lambda_k (\mu_{k,\ell} A_{j_{(k,\ell,1)}} \cdots A_{j_{(k,\ell,N)}}) A_{i_{(k,N)}},$$

which implies $B \in A^{N+1}$. Since B is arbitrary, we conclude that $A^{N+1} = (\mathbb{C}^D)^{\otimes 2}$. $\square$

We are now ready to define the injectivity radius.

Definition 2.3.2. The *injectivity radius* is the smallest integer $C_{D,d}$ such that for every *spanning* $A \in V$ we have $i(A) \leq C_{D,d}$. Furthermore, we write $C_D = C_{D,D^2} = \max_d C_{D,d}$.

Remark 2.3.3. The name "injectivity radius" stems from the fact that $N \geq i(A)$ if and only if the map

$$\mathbb{C}^{D \times D} \to (\mathbb{C}^d)^N$$

$$B \mapsto \sum_{i_1,\ldots,i_N} \mathrm{tr}(B \cdot A_{i_1} \cdots A_{i_N}) e_{i_1} \otimes \cdots \otimes e_{i_N}$$

is injective. See [PGVWC07, 3.2.4] for more details.

From the definition, it is not clear that a finite injectivity radius exists for every (D, d). The existence of a finite injectivity radius was first established by Sanz et al. [SPGWC10], by proving that for every spanning $A \in V$, it holds that $i(A) \leq (D^2 - \dim \langle A \rangle + 1)D^2$. In particular, $C_D = O(D^4)$. More recently, this bound was improved to $C_D = O(D^2 \log D)$ (more precisely: $C_D \leq 2D^2(6 + \log_2 D)$) by Michalek and Shitov [MS19]. It is conjectured that $C_D = O(D^2)$ [PGVWC07, Conjecture 2].

Remark 2.3.4. A bound on $C_{D,d}$ or C_D is often referred to as a *quantum Wielandt inequality*. The classical Wielandt inequality, first stated in [Wie50], is a statement about matrices with nonnegative entries. In our language, we can phrase it as follows: "If $A \in V$ is spanning *and every A_i has at most one nonzero entry*, then $i(A) \leq D^2 - 2D + 2$." This bound is known to be sharp.

Remark 2.3.5. One naive strategy of proving $C_D = O(D^2)$ would be to show that for $A \in V$ spanning, it holds that $\dim A^{L+1} > \dim A^L$ whenever $A^L \subsetneq (\mathbb{C}^D)^{\otimes 2}$. However, this stronger statement is false. For instance, let $D = 2m + 2$ and let $A^1 = \langle A \rangle$ be the space of $D \times D$ matrices of the form

$$\begin{pmatrix} a & v^T & 0 & 0 \\ 0 & 0 & B & 0 \\ 0 & 0 & 0 & w \\ b & 0 & 0 & 0 \end{pmatrix},$$

where v and w are $m \times 1$-matrices and B is an $m \times m$-matrix. Then A is spanning since $A^4 = (\mathbb{C}^D)^{\otimes 2}$, but A^2 only contains the matrices of the form

$$\left(\begin{array}{c|c|c|c} a & v_1^T & v_2^T & 0 \\[1ex] \hline 0 & 0 & 0 & w_1 \\[1ex] \hline w_2 & 0 & 0 & 0 \\[1ex] \hline b & v_3^T & 0 & 0 \end{array} \right).$$

Hence $\dim(A^2) = 5m + 2 \leq m^2 + 2m + 2 = \dim(A^1)$, assuming we chose $m \geq 3$.

All the bounds on C_D mentioned above were proven by explicit methods from linear algebra. Below, we present a very short, but nonconstructive, proof for the existence of a finite injectivity radius. We will need the following lemma.

Lemma 2.3.6. *The map $V \to \mathbb{N} : A \mapsto i(A)$ is upper semicontinuous: for every $N \in \mathbb{N}$, the set $V_N := \{A \in V : i(A) \geq N\}$ is Zariski closed.*

Proof. For every $A \in V$, we consider the $d^N \times D^2$ matrix $M_N(A)$ whose rows consist of the entries of the matrices $A_{i_1} \cdots A_{i_N}$. Note that the entries of $M_N(A)$ are polynomials in the entries of A. Now, by Lemma 2.3.1, the condition $i(A) > N$ is equivalent to $A^N \subsetneq (\mathbb{C}^D)^{\otimes 2}$, which is equivalent to the matrix $M_N(A)$ having rank smaller than D^2. This is equivalent to the maximal minors of $M_N(A)$ vanishing, which is a polynomial condition in the entries of A. Hence V_N is Zariski closed. $\qquad\square$

The existence of a finite injectivity radius now becomes a simple consequence of Hilbert's Basis Theorem.

Theorem 2.3.7 (Quantum Wielandt theorem). *For every D, d, there exists a $C_{D,d}$ such that for every spanning $A \in V = (\mathbb{C}^D)^{\otimes 2} \otimes \mathbb{C}^d$, already $A^{C_{D,d}} = (\mathbb{C}^D)^{\otimes 2}$.*

Proof. Consider the chain $V_1 \supseteq \ldots \supseteq V_N \supseteq \ldots$, where $V_N := \{A \in V : i(A) \geq N\}$. By Lemma 2.3.6, V_N is Zariski closed. Hence by Hilbert's Basis Theorem, there is a constant C such that $V_N = V_C$ for all $C \geq N$. This C is our desired constant: if $A \in V_C$ then A is not spanning; if $A \notin V_C$ then $A^C = (\mathbb{C}^D)^{\otimes 2}$. $\qquad\square$

Our proof of Theorem 2.3.7 is shorter and less explicit then the proofs that were known before, but gives a weaker, nonconstructive result. However, it turns out that our proof technique can be generalized to a setting where (thus far) constructive methods have failed, yielding a *tensor version of the quantum Wielandt theorem*. This will be the main topic of Section 2.5. We now turn our attention to the generic injectivity radius.

Definition 2.3.8. The *generic injectivity radius* $GC_{D,d}$ is the spanning index $i(A)$ of a *generic* $A \in V$.

By Lemma 2.3.6, $GC_{D,d}$ is equal to $\min\{i(A)|A \in V\}$. The generic injectivity radius is much easier to control, as to bound it from above it is enough to exhibit one tuple of matrices that generate the whole space fast. It is clear that $GC_{D,d} \geq \log_d(D^2)$, as $\dim L^N \leq d^N$. The following result by Klep and Špenko solves the problem of determining $GC_{D,d}$ almost completely.

Theorem 2.3.9 ([KŠ16, Corollary 2.4]). *For each d satisfying $D^2 \leq d^{2m}$, there exists a set of d matrices such that words of length $2m$ in the matrices span $\mathbb{C}^{D \times D}$.*

Corollary 2.3.10. *If $\lceil 2\log_d(D) \rceil$ is even, then it holds that $GC_{D,d} = \lceil 2\log_d(D) \rceil$. If $\lceil 2\log_d(D) \rceil$ is odd, then $GC_{D,d}$ is either $\lceil 2\log_d(D) \rceil$ or $\lceil 2\log_d(D) \rceil + 1$. In particular, $GC_{D,d} = O(\log D)$.*

2.4 Uniform matrix product states: geometric properties

This section is devoted to our main results regarding the structure of the set $\mathrm{uMPS}(D, d, N)$ of uniform matrix product states, and its Zariski (or Euclidean) closure $\overline{\mathrm{uMPS}(D, d, N)}$.

2.4.1 Topological properties

We start by giving a complete classification when $\mathrm{uMPS}(2, d, N)$ is a closed set.

Theorem 2.4.1. *For $d > 1$ and $N > 2$, $\mathrm{uMPS}(2, d, N)$ is closed if and only if $(d, N) = (2, 3)$.*

Proof. By 2.2.6, it suffices to show that $\mathrm{uMPS}(2, 2, 3)$ is closed, that $\mathrm{uMPS}(2, 3, 3)$ is not closed, and that $\mathrm{uMPS}(2, 2, N)$ is not closed if $N > 3$.

Step 1. The set $\mathrm{uMPS}(2, 2, 3)$ is closed because it equals the ambient space $\mathrm{Cyc}^3(\mathbb{C}^2) = \mathrm{Sym}^3(\mathbb{C}^2)$. This was already proven in [HMS19, Section 5.2], so we only sketch the proof here: by Remark 2.2.14, it suffices to find one tensor in every GL_2-orbit that is in $\mathrm{uMPS}(2, 2, 3)$. Since there are only three GL_2-orbits, this can be done explicitly.

Step 2. Next, we show that $\mathrm{uMPS}(2, 3, 3)$ is not closed. To do this, it suffices to construct a tensor T, which is in $\overline{\mathrm{uMPS}(2, 3, 3)}$, but not in $\mathrm{uMPS}(2, 3, 3)$. We will prove that the following tensor works:

$$T = e_{012} = e_0 \otimes e_1 \otimes e_2 + e_1 \otimes e_2 \otimes e_0 + e_2 \otimes e_0 \otimes e_1 \in \mathrm{Cyc}^3(\mathbb{C}^3).$$

Step 2a. First we show that $T \in \overline{\mathrm{uMPS}(2,3,3)}$ by exhibiting T as a limit of tensors in $\mathrm{uMPS}(2,3,3)$. Consider for every $\lambda \in \mathbb{C} \setminus \{0\}$ the following three matrices:

$$M_{0,\lambda} = \lambda^2 \begin{pmatrix} 1 & 0 \\ 0 & 0 \end{pmatrix}, \quad M_{1,\lambda} = \lambda^{-1} \begin{pmatrix} 0 & 1 \\ 0 & 0 \end{pmatrix}, \quad M_{2,\lambda} = \lambda^{-1} \begin{pmatrix} 0 & 0 \\ 1 & 0 \end{pmatrix}.$$

Then by definition

$$T_3(M_{0,\lambda}, M_{1,\lambda}, M_{2,\lambda}) = \lambda^6 e_{000} + e_{012},$$

so we have $T = \lim_{\lambda \to 0} T_3(M_{0,\lambda}, M_{1,\lambda}, M_{2,\lambda})$, hence $T \in \overline{\mathrm{uMPS}(2,3,3)}$.

Step 2b. Proving that $T \notin \mathrm{uMPS}(2,3,3)$ amounts to showing that a certain system of polynomial equations has no solutions. We assume that $T \in \mathrm{uMPS}(2,3,3)$ and derive a contradiction. We may write

$$T = e_{012} = T_3(M_0, M_1, M_2)$$

for some $M_0, M_1, M_2 \in \mathbb{C}^{2\times 2}$, where M_0 is in Jordan normal form. Suppose first that $M_0 = \begin{pmatrix} a & 1 \\ 0 & a \end{pmatrix}$. Since $\mathrm{tr}(M_0^3) = 0$, we get $a = 0$. Let us furthermore write $M_1 = \begin{pmatrix} a_2 & b_2 \\ c_2 & d_2 \end{pmatrix}$ and $M_2 = \begin{pmatrix} a_3 & b_3 \\ c_3 & d_3 \end{pmatrix}$.
Then since $\mathrm{tr}(M_0 M_1^2) = \mathrm{tr}(M_0 M_2^2) = \mathrm{tr}(M_0 M_2 M_1) = 0$ and $\mathrm{tr}(M_0 M_1 M_2) = 1$, we get that $(a_2, b_2, c_2, d_2, a_3, b_3, c_3, d_3)$ must be a solution of the following system of four equations:

$$\begin{cases} c_2(a_2 + d_2) = 0 \\ c_3(a_3 + d_3) = 0 \\ c_3 a_2 + c_2 d_3 = 0 \\ c_2 a_3 + c_3 d_2 = 1 \end{cases}$$

It is not hard to see that this system has no solutions.

Suppose now that $M_0 = \begin{pmatrix} a_1 & 0 \\ 0 & d_1 \end{pmatrix}$, and write $M_1 = \begin{pmatrix} a_2 & b_2 \\ c_2 & d_2 \end{pmatrix}$ and $M_2 = \begin{pmatrix} a_3 & b_3 \\ c_3 & d_3 \end{pmatrix}$ as before. Then $(a_1, d_1, a_2, b_2, c_2, d_2, a_3, b_3, c_3, d_3)$ must be a solution of the following system:

$$\begin{cases} a_1^3 + d_1^3 = 0 \\ a_2^3 + 3a_2b_2c_2 + 3b_2c_2d_2 + d_2^3 = 0 \\ a_3^3 + 3a_3b_3c_3 + 3b_3c_3d_3 + d_3^3 = 0 \\ a_1a_2^2 + a_1b_2c_2 + b_2c_2d_1 + d_1d_2^2 = 0 \\ a_1^2a_2 + d_1^2d_2 = 0 \\ a_1a_3^2 + a_1b_3c_3 + b_3c_3d_1 + d_1d_3^2 = 0 \\ a_1^2a_3 + d_1^2d_3 = 0 \\ a_2a_3^2 + a_3b_3c_2 + a_3b_2c_3 + a_2b_3c_3 + b_3c_3d_2 + b_3c_2d_3 + b_2c_3d_3 + d_2d_3^2 = 0 \\ a_2^2a_3 + a_3b_2c_2 + a_2b_3c_2 + a_2b_2c_3 + b_3c_2d_2 + b_2c_3d_2 + b_2c_2d_3 + d_2^2d_3 = 0 \\ a_1a_2a_3 + a_1b_2c_3 + b_3c_2d_1 + d_1d_2d_3 = 1 \\ a_1a_2a_3 + a_1b_3c_2 + b_2c_3d_1 + d_1d_2d_3 = 0 \end{cases}$$

One can show that this system has no solutions for example by computing a Gröbner basis in Macaulay2. This leads to a contradiction.

Step 3. Next, we need to show that $\text{uMPS}(2,2,N)$ is not closed for any $N > 3$. We will do this by showing that the *W-state* W_N from Definition 2.2.15 is in $\overline{\text{uMPS}(2,2,N)}$, but not in $\text{uMPS}(2,2,N)$.

Step 3a. We can exhibit the *W-state* W_N from Definition 2.2.15 as a limit of tensors in $\text{uMPS}(2,2,N)$. Let ζ be any complex number satisfying $\zeta^N = -1$. For every $\lambda \neq 0$, consider the matrices $M_{0,\lambda} = \lambda^{-1}\begin{pmatrix} 1 & 0 \\ 0 & \zeta \end{pmatrix}$ and $M_{1,\lambda} = \lambda^{N-1}\begin{pmatrix} 1 & 0 \\ 0 & -\zeta \end{pmatrix}$. Then it is easy to see that

$$T_N(M_{0,\lambda}, M_{1,\lambda}) = 2W_N + O(\lambda),$$

where $O(\lambda)$ means higher order terms in λ. Letting $\lambda \to 0$ shows that $W_N \in \overline{\text{uMPS}(2,2,N)}$.

Step 3b. Showing that $W_N \notin \text{uMPS}(2,2,N)$ is done by an explicit computational argument. Let us write $T = e_{0\ldots01}$ and assume that $T \in \text{uMPS}(2,2,N)$. Then we can write $T = T_N(M_0, M_1)$, for some 2×2-matrices M_0 and M_1, where M_0 is in Jordan normal form.

First suppose $M_0 = \begin{pmatrix} a & 1 \\ 0 & a \end{pmatrix}$. Then since $\text{tr}(M_0^N) = 0$, we get $a = 0$. But then also $\text{tr}(M_0^{N-1}M_1) = 0$. This is a contradiction with our assumption that $\text{tr}(M_0^{N-1}M_1) = 1$.

Thus, we can assume that $M_0 = \begin{pmatrix} a & 0 \\ 0 & d \end{pmatrix}$ and write $M_1 = \begin{pmatrix} A & B \\ C & D \end{pmatrix}$.

First note that since $\text{tr}(M_0^N) = 0$, we get that $a = \zeta d$, where $\zeta^N = -1$. It is clear that $a \neq 0$ and $d \neq 0$, since otherwise $M_0 = 0$.

The equation $\operatorname{tr}(M_0^{N-1}M_1) = 1$ becomes

$$a^{N-1}A + d^{N-1}D = 1$$
$$\iff a^{N-1}A + a^{N-1}\zeta^{N-1}D = 1$$
$$\implies A + \zeta^{N-1}D \neq 0. \tag{2.4.1}$$

We also get that, for every $s \in \{0, 1, \ldots N-2\}$:

$$\operatorname{tr}(M_0^s M_1 M_0^{N-2-s} M_1) = 0$$
$$\iff a^{N-2}A^2 + (a^s d^{N-2-s} + a^{N-2-s}d^s)BC + d^{N-2}D^2 = 0$$
$$\iff A^2 + (\zeta^s + \zeta^{N-2-s})BC + \zeta^{N-2}D^2 = 0. \tag{2.4.2}$$

Furthermore, for every $t \in \{0, 1, \ldots N-3\}$:

$$\operatorname{tr}(M_0^t M_1^2 M_0^{N-3-t} M_1) = 0$$
$$\iff A(A^2 + BC) + (\zeta^t + \zeta^{N-3-t})BC(A + D) + \zeta^{N-3}D(D^2 + BC) = 0. \tag{2.4.3}$$

We now show that the equalities (2.4.2) and (2.4.3), together with the inequality (2.4.1), lead to a contradiction. The proof is not hard, but we need to distinguish some cases. In the proof it will turn out that we only need (2.4.2) for $s \in \{0, 1, 2\}$, and (2.4.3) for $t \in \{0, 1\}$.

Case 1. $BC = 0$.

Then we get that

$$A + \zeta^{N-1}D \neq 0$$
$$A^2 + \zeta^{N-2}D^2 = 0$$
$$A^3 + \zeta^{N-3}D^3 = 0$$

Hence $\zeta^{N-2}AD^2 = -A^3 = \zeta^{N-3}D^3$, which implies $D^2(\zeta A - D)$, but this leads to a contradiction: either $\zeta A - D = 0$, but this is a contradiction with (2.4.1); or $D = 0$, which implies $A = 0$ hence also yields a contradiction with (2.4.1).

Case 2. $BC \neq 0$.

Then (2.4.2) tells us that for every $s \in \{0, 1, \ldots N-2\}$, it holds that $1 + \zeta^{N-2} = \zeta^s + \zeta^{N-2-s}$, so $(1 - \zeta^s)(1 - \zeta^{N-2-s}) = 0$. Putting $s = 1$ yields $\zeta^{N-3} = 1$. Putting $s = 2$ yields $\zeta^{N-4} = 1$ or $\zeta^2 = 1$. The former would imply $\zeta = 1$, a contradiction. So we get $\zeta = -1$ (and N odd). Note that here we used that $N \geq 4$. Now (2.4.1) tells us that $A + D \neq 0$. But then by (2.4.3) for $t = 0$ and $t = 1$, we get $1 + \zeta^{N-3} = \zeta + \zeta^{N-2}$. Since $\zeta = -1$ and N is odd, that is a contradiction.

$$\square$$

Remark 2.4.2. Uniform matrix product states can also be defined over other fields; the definition remains the same. The statement of Theorem 2.4.1 is also true when working over $\mathbb{R}$ instead of $\mathbb{C}$. The only additional things we need to show for this are:

- Also over $\mathbb{R}$, it holds that $\mathrm{uMPS}(D, d, N) = \mathrm{Sym}^3(\mathbb{R}^2)$. The proof is analogous to the complex case (this time there are four GL_2-orbits).

- The W-state is also in the closure of real-valued $\mathrm{uMPS}(D, d, N)$. This can be achieved by replacing the complex matrices $M_{0,\lambda}$, $M_{1,\lambda}$ in the proof by the real matrices $M_{0,\lambda} = \lambda^{-1} \begin{pmatrix} \zeta + \zeta^{-1} & 1 \\ -1 & 0 \end{pmatrix}$ and $M_{1,\lambda} = \lambda^{N-1} \begin{pmatrix} 1 & 0 \\ 0 & 1 \end{pmatrix}$.

We proceed to show that for any fixed D, d, $\mathrm{uMPS}(D, d, N)$ will not be closed for large N. This is particularly important in quantum physics, where N is typically assumed to be very large. Our main ingredient is the following Theorem by Perez-Garcia et al.

Theorem 2.4.3 (See [PGVWC07, Corollary 1]). *When $N > 6(D - 1)(C_D + 1)$, we have that $W_N \notin \mathrm{uMPS}(D, 2, N)$.*

Corollary 2.4.4. *If $N > 6(D - 1)(C_D + 1)$, then $\mathrm{uMPS}(D, d, N)$ is not closed.*

Proof. It suffices to consider the case $d = 2$. Our arguments from the proof of Theorem 2.4.1 show that $W_N \in \overline{\mathrm{uMPS}(D, 2, N)}$ for every D. We conclude by Theorem 2.4.3. $\qquad\square$

Conjecture 2.4.5. *Except for the trivial cases in Proposition 2.2.7, the set $\mathrm{uMPS}(D, d, N)$ is only closed if it fills the ambient space $\mathrm{Cyc}^N(\mathbb{C}^d)$.*

We now discuss connectedness. We will do this both in the real and the complex case, so for the rest of this section, let $\mathbb{K} = \mathbb{C}$ or $\mathbb{R}$. The set $\mathrm{uMPS}(D, d, N)$ is clearly connected, since every tensor in it can be rescaled to 0. We will instead consider the projectivization of $\mathrm{uMPS}(D, d, N)$. This is the image of a rational map

$$T_N : \mathbb{P}((\mathbb{K}^{D \times D})^d) \dashrightarrow \mathbb{P}((\mathbb{K}^d)^{\otimes N}).$$

We recall that a rational map is in general only defined on an open subset; the points where it is not defined (i.e. all defining polynomials vanish) form a closed subvariety called the *base locus*. Our result about connectedness will follow from the following lemma.

Lemma 2.4.6. *Let $F : \mathbb{P}^n \dashrightarrow \mathbb{P}^N$ be a rational map. Then we have the following:*

1. over $\mathbb{C}$ the image is always connected.

2. over $\mathbb{R}$ the image is connected if the codimension of the base locus is greater than one.

Proof. It suffices to show that the complement of the base locus is connected. In the case $\mathbb{K} = \mathbb{C}$ this is immediate, since the base locus is a subvariety of $\mathbb{P}^n$. In the case $\mathbb{K} = \mathbb{R}$ it follows from the assumption on the codimension. $\qquad\square$

Theorem 2.4.7. *Both over $\mathbb{C}$ and over $\mathbb{R}$, for any choice of the parameters, the projectivization of $\mathrm{uMPS}(D, d, N)$ is connected.*

Proof. The base locus $Z \subseteq \mathbb{P}((\mathbb{K}^{D \times D})^d)$ of T_N consists af all $(M_1, \dots, M_d)$ such that for every choice of indices $\mathrm{tr}(M_{i_1} \cdots M_{i_N}) = 0$. By Lemma 2.4.6, we only need to check that Z is not a hypersurface. But this is obvious: if Z were a hypersurface, all polynomials $\mathrm{tr}(M_{i_1} \cdots M_{i_N})$ would be divisible by the same polynomial. This is clearly not the case, as $\mathrm{tr}(M_1^N)$ and $\mathrm{tr}(M_2^N)$ do not share any variables. $\qquad\square$

Problem 2.4.8. What can be said about the higher homotopy and homology of $\mathrm{uMPS}(D, d, N)$ and its projectivization?

2.4.2 Surjectivity

In this section we study for which parameters the set $\mathrm{uMPS}(D, d, N)$ fills the ambient space $\mathrm{Cyc}^N(\mathbb{C}^d)$. In the first part we investigate if $\mathrm{uMPS}(D, d, N)$ can be contained in a linear subspace of $\mathrm{Cyc}^N(\mathbb{C}^d)$. In the second part we prove that, for fixed d and N, $\mathrm{uMPS}(D, d, N) = \mathrm{Cyc}^N(\mathbb{C}^d)$ if D is large enough. More precisely, it suffices to take $D \geq N \cdot \dim(\mathrm{Cyc}^N(\mathbb{C}^d))$ (see Corollary 2.4.16). In the last part we describe a very useful surjectivity criterion for polynomial maps, and apply it to the case $(D, d, N) = (3, 2, 4)$. The following result slightly generalizes [GLW18, Proposition 3.1].

Proposition 2.4.9. *If $D \geq N$, the linear span of $\mathrm{uMPS}(D, d, N)$ is equal to the whole space $\mathrm{Cyc}^N(\mathbb{C}^d)$.*

Proof. The case $d \geq N$ was proven in [GLW18, Proposition 3.1].
Suppose $d < N$. Then the projection $(\mathbb{C}^N)^{\otimes N} \to (\mathbb{C}^d)^{\otimes N}$ maps $\mathrm{uMPS}(D, N, N)$ to $\mathrm{uMPS}(D, d, N)$ as in Lemma 2.2.5. The proposition follows. $\qquad\square$

Now we will analyze how Proposition 2.4.9 fails if we drop the assumption $D \geq N$. We start with a trivial example.

Example 2.4.10. If $D = 1$ then $\langle \mathrm{uMPS}(1, d, N) \rangle \subseteq \mathrm{Sym}^N(\mathbb{C}^d)$, which is a strict subspace of $\mathrm{Cyc}^N(\mathbb{C}^d)$ unless $N \leq 2$, $d \leq 1$, or $(d, N) = (2, 3)$.

More surprisingly, even for $D > 1$ it can still happen that $\mathrm{uMPS}(D, d, N)$ is contained in a strict linear subspace of $\mathrm{Cyc}^N(\mathbb{C}^d)$, as the example below shows.

Example 2.4.11. If M_0, M_1 are two 2×2 matrices, then for any sequence $(i_1, \ldots, i_N) \in \{0, 1\}^N$, it holds that $\mathrm{tr}(M_{i_1} M_{i_2} \ldots M_{i_N}) = \mathrm{tr}(M_{i_N} M_{i_{N-1}} \ldots M_{i_1})$. This can be shown inductively on N using the Cayley-Hamilton relation $M^2 = \mathrm{tr}(M)M - \det(M)I$ for 2×2 matrices, see [Gre14, Theorem 1.1]. As a corollary, the set $\mathrm{uMPS}(2, 2, N)$ is contained in a strict linear subspace of $\mathrm{Cyc}^N(\mathbb{C}^2)$ for all $N \geq 6$. For instance, for every pair of 2×2-matrices M_0 and M_1, it holds that $\mathrm{tr}(M_0^2 M_1^2 M_0 M_1) = \mathrm{tr}(M_1 M_0 M_1^2 M_0^2)$, even though $(0, 0, 1, 1, 0, 1)$ and $(1, 0, 1, 1, 0, 0)$ are not the same up to cyclic permutation. Hence $\mathrm{Cyc}^6(\mathbb{C}^2)$ is contained in the linear subspace of $\mathrm{Cyc}^N(\mathbb{C}^2)$ defined by $x_{001101} = x_{101100}$. We stress that this "reflection symmetry" is specific for the case $D = d = 2$. For $N = 8$, we found linear relations that don't follow from the reflection symmetry.

In the following theorem, let $C(N_0, N_1)$ denote the number of sequences consisting of N_0 times the symbol '0' and N_1 times the symbol '1', where we identify two sequences if they are the same up to cyclic permutation.

Theorem 2.4.12. *If*

$$C(N_0, N_1) > \binom{N_0 + D - 1}{D - 1} \binom{N_1 + D^2 - D}{D^2 - D}$$

then for every $d \geq 2$, $\mathrm{uMPS}(D, d, N_0 + N_1)$ is contained in a strict linear subspace of $\mathrm{Cyc}^{N_0 + N_1}(\mathbb{C}^d)$.

Proof. It clearly suffices to show the theorem for d=2. As in the proof of Proposition 2.2.9, $\overline{\mathrm{uMPS}(D, d, N)}$ is the closure of the image of a polynomial map $\phi : Z \to \mathrm{Cyc}^N(\mathbb{C}^2)$, where Z is the $D^2 + 1$-dimensional space of pairs of $D \times D$-matrices (M_0, M_1) for which M_0 is diagonal and the $D - 1$ nondiagonal entries on the top row of M_1 are all equal to 1. Consider the set of all polynomials $\mathrm{tr}(M_{i_1} \cdots M_{i_N})$, where exactly N_0 of the indices i_j are 0 and the other N_1 indices are 1. These polynomials have degree N_0 in the first D variables, and degree N_1 in the last $D^2 - D + 1$ variables. The space of such polynomials has dimension $\binom{N_0 + D - 1}{D - 1} \binom{N_1 + D^2 - D}{D^2 - D}$. Hence, by the assumption, some of these polynomials must be linearly dependent. This imposes a linear condition on the image of ϕ. $\qquad\square$

Remark 2.4.13. For large N_0, N_1 the assumptions of the previous theorem happen very often, as the left hand side grows exponentially, while the right hand side grows polynomially.

Our next goal is to show that $\mathrm{uMPS}(D, d, N)$ fills the ambient space for large D. For $X_1, \ldots, X_m$ subsets of a vector space V, $\mathrm{join}(X_1, \ldots, X_m)$ will denote their join $\bigcup_{x_i \in X_i} \langle x_1, \ldots, x_m \rangle$.

Lemma 2.4.14. *Let $X_i = \mathrm{uMPS}(D_i, d, N)$ for $1 \leq i \leq m$. Then*

$$\mathrm{join}(X_1, \ldots, X_m) \subseteq \mathrm{uMPS}(\sum D_i, d, N).$$

Proof. Let $v \in \mathrm{join}(X_1, \ldots, X_m)$. Then $v = \sum_{i=1}^{m} v_i$, where $v_i \in \mathrm{uMPS}(D_i, d, N)$. There exist $D_i \times D_i$-matrices $M_{i,j}$ such that $v_i = T_N(M_{i,0}, \ldots, M_{i,d-1})$. Thus

$$
\begin{aligned}
v &= \sum_{i=1}^{m} T_N(M_{i,0}, \ldots, M_{i,d-1}) \\
&= T_N\Big(\mathrm{diag}(M_{1,0}, \ldots, M_{m,0}), \ldots, \mathrm{diag}(M_{1,d-1}, \ldots, M_{m,d-1}) \Big) \\
&\in \mathrm{uMPS}(\textstyle\sum D_i, d, N).
\end{aligned}
$$

$\square$

Proposition 2.4.15. *Suppose that the linear span of* $\mathrm{uMPS}(D, d, N)$ *is equal to* $\mathrm{Cyc}^N(\mathbb{C}^d)$. *Then* $\mathrm{uMPS}(D \cdot \dim(\mathrm{Cyc}^N(\mathbb{C}^d)), d, N) = \mathrm{Cyc}^N(\mathbb{C}^d)$.

Proof. Writing $m = \dim(\mathrm{Cyc}^N(\mathbb{C}^d))$, by assumption $\mathrm{Cyc}^N(\mathbb{C}^d)$ is equal to the join of m copies of $\mathrm{uMPS}(D, d, N)$. The result now follows from Lemma 2.4.14. $\square$

Corollary 2.4.16. $\mathrm{uMPS}(D, d, N) = \mathrm{Cyc}^N(\mathbb{C}^d)$ *for* $D \geq N \cdot \dim(\mathrm{Cyc}^N(\mathbb{C}^d))$.

Proof. Follows immediately from Propostitions 2.4.9 and 2.4.15. $\square$

Remark 2.4.17. Checking whether $\overline{\mathrm{uMPS}(D, d, N)}$ fills the ambient space is a computationally easy task for small parameter values. Indeed, since the set $\mathrm{uMPS}(D, d, N)$ is a constructible subset, its closure fills the ambient space if and only if it is full-dimensional. In particular for all cases in Table 1, $\overline{\mathrm{uMPS}(D, d, N)}$ fills the ambient space if and only if $(d-1)D^2 + 1 \geq \dim(\mathrm{Cyc}^N(\mathbb{C}^d))$ (see also Remark 2.2.12)). However, checking whether $\mathrm{uMPS}(D, d, N)$ fills the ambient space for fixed parameter values is a significantly more difficult task, which we address now.

In this part we describe a general criterion for $\mathrm{uMPS}(D, d, N)$ to fill the space, and apply it to show that $\mathrm{uMPS}(3, 2, 4)$ fills the space.
Let $f \colon \mathbb{C}^n \to \mathbb{C}^m$ be a polynomial map defined by $x \mapsto (f_1(x), \ldots, f_m(x))$, where the f_i are homogeneous polynomials of the same degree in the coordinates of $x = (x_1, \ldots, x_n)$. Instead of focusing on the above map f, we want to consider the rational projective map $\mathbb{P}^{n-1} \dashrightarrow \mathbb{P}^{m-1}$. We prove the following theorem.

Theorem 2.4.18. *Let* $f \colon \mathbb{P}^{n-1} \dashrightarrow \mathbb{P}^{m-1}$ *be a rational projective map and* B *be the base locus of* f. *If there exists a subspace* $Y \subset \mathbb{P}^{n-1}$ *of dimension* $\mathrm{im}(f)$, *which is disjoint with the base locus* B, *then the map* f *has closed image. In particular: if*

$$
\dim(B) + \dim(\mathrm{im}(f)) < n - 1,
$$

then the map f *has closed image.*

Proof. Let d be the dimension of the image $\mathrm{im}(f)$ and b the dimension of the base locus B. Then all non-empty fibers of f have at least dimension $n-1-d$, so Y intersects every fiber. Hence

$$\mathrm{im}(f) = \mathrm{im}(f|_Y).$$

However, since $Y \cap B = \emptyset$, f is well-defined on Y, which is compact, and hence the image is closed. In particular, if $b + d < n - 1$, then a generic d-dimensional subspace Y will be disjoint with the base locus, so that f has closed image by the above argument. $\qquad\square$

Using the above theorem we can deduce that for showing the surjectivity of f it is enough to find a sufficiently big space on which the map f is well defined.

Example 2.4.19. Let us apply Theorem 2.4.18 to the case $(D, d, N) = (3, 2, 4)$. It is computationally difficult to compute the dimension of the base locus B, but to show that f is surjective it is enough to find a $\mathbb{P}^5 \subset \mathbb{P}^{17}$, disjoint with B. Below we present the Macaulay2 code providing the $\mathbb{P}^5$ which satisfies the above assumptions.

```
R=QQ[a_1,b_1,c_1,c_2,e_2,h_2];
M1=matrix{{a_1,b_1,c_1},{c_2,e_2,h_2},
{c_1+b_1-3*c_2,h_2-e_2,2*a_1-7*b_1}};
M2=matrix{{a_1+2*c_2,e_2+5*h_2,-c_1-3*e_2},
{b_1+a_1-2*h_2,e_2-5*c_2,b_1-c_1+13*a_1},
{h_2-b_1-c_1,c_2+3*a_1-2*e_2,a_1-h_2}};
a=trace(M1*M1*M1*M1);
b=trace(M1*M1*M1*M2);
c=trace(M1*M1*M2*M2);
d=trace(M1*M2*M1*M2);
e=trace(M1*M2*M2*M2);
f=trace(M2*M2*M2*M2);
I=ideal(a,b,c,d,e,f);
(dim I) == 0
```

Remark 2.4.20. We now explain how to find the given $\mathbb{P}^5$. Taking a completely general $\mathbb{P}^5$, although from a theoretical point of view most desirable, is not possible due to computational restraints. On the other hand taking very special, simple $\mathbb{P}^5$ usually leads to intersection with B that is of large dimension. The given $\mathbb{P}^5$ was found by first considering a special $\mathbb{P}^5$ and computing the dimension and degree of the intersection. The $\mathbb{P}^5$ was successively modified to a more general one, each time computing the dimension and degree. The degree (in most cases) or dimension of the intersection were dropping, while we modified the $\mathbb{P}^5$. This meant that we were not in a generic situation and further modifications were possible. Finally, we reached the given example.

2.4.3 The trace parametrization

In this section we describe another parametrization of the uniform matrix product states. Let us consider a d-tuple $(M_0, \ldots, M_{d-1})$ of $D \times D$-matrices with indeterminate entries. The *trace algebra* $\mathcal{C}_{D,d}$ is the algebra generated by the traces of products $\mathrm{tr}(M_{i_1} \cdots M_{i_k})$, seen as polynomials in the entries of the M_i. It can alternatively be described as follows: if R is the polynomial ring in the entries of the M_i, then $\mathcal{C}_{D,d}$ is isomorphic to the invariant ring R^{GL_D}, where the group action comes from simultaneously conjugating the matrices [Sib68, Ler76, Pro76].

It follows from a standard fact in invariant theory [Dol03, Prop. 3.1] that the trace algebra is generated by finitely many traces $T_1, \ldots, T_K$, where every T_j is an expression of the form $\mathrm{tr}(M_{i_1} \cdots M_{i_k})$. We conclude that $\mathcal{C}_{D,d}$ is isomorphic to a quotient algebra $\mathbb{C}[T_1, \ldots, T_K]/(f_1, \ldots, f_r)$ for some polynomials $f_i(T_1, \ldots, T_K)$. We will now parametrize $\mathrm{uMPS}(D, d, N)$ with the *spectrum* of the algebra $\mathcal{C}_{D,d}$. For readers not familiar with the Spec construction: $\mathrm{Spec}(\mathcal{C}_{D,d})$ may be regarded as the set of all K-tuples $(t_1, \ldots, t_K) \in \mathbb{C}^K$ satisfying the equations $f_i(t_1, \ldots, t_K) = 0$. We have the following diagram:

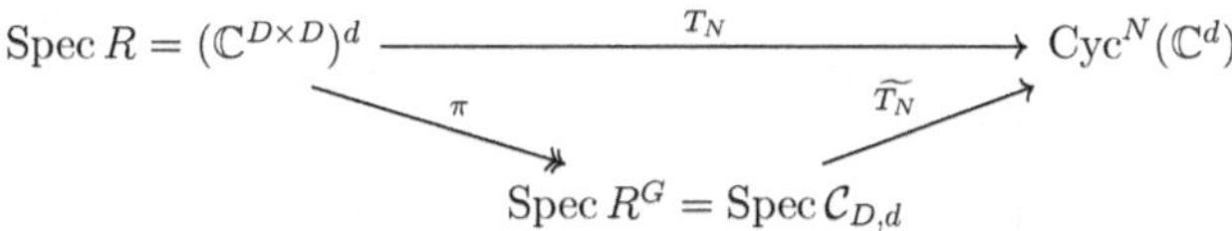

The map $\pi : (\mathbb{C}^{D \times D})^d \to \mathrm{Spec}\,\mathcal{C}_{D,d}$ is a good categorical quotient [Dol03, Thm. 6.1], in particular π is surjective. Hence, the maps T_N and $\widetilde{T_N}$ have the same image $\mathrm{uMPS}(D, d, N)$. The reader not familiar with categorical and geometric quotients can consult [Dol03, Section 6.1] or [CLS11, Section 5.0] for definitions and basic properties.

Sibirskii [Sib68] showed in the case $D = 2$ that the trace algebra $\mathcal{C}_{2,d}$ is minimally generated by the elements $\mathrm{tr}(M_i)$, $\mathrm{tr}(M_i M_j)$ for $0 \leq i \leq j \leq d-1$, and $\mathrm{tr}(M_i M_j M_k)$ for $0 \leq i < j < k \leq d-1$. Moreover in the case $D = d = 2$, there are no relations between the five generators $\mathrm{tr}(M_0)$, $\mathrm{tr}(M_1)$, $\mathrm{tr}(M_0^2)$, $\mathrm{tr}(M_0 M_1)$, $\mathrm{tr}(M_1^2)$. In other words, $\mathcal{C}_{2,2}$ is the polynomial ring in 5 variables. This means that we get a parametrization $\widetilde{T_N} \colon \mathbb{C}^5 \to \mathrm{uMPS}(2, 2, N)$.

Using the trace parametrization and Macaulay2, it is possible to obtain equations for $\mathrm{uMPS}(2, 2, N)$ for small values of N.

Theorem 2.4.21. *1. [CM14, Theorem 3] The ideal of $\mathrm{uMPS}(2, 2, 4) \subset \mathbb{C}^6$ is generated by one sextic.*

2. [CM14, Question after Theorem 4] The ideal of $\mathrm{uMPS}(2, 2, 5) \subset \mathbb{C}^8$ is generated by 3 quartics and 27 sextics.

3. *The ideal of* uMPS$(2, 2, 6) \subset \mathbb{C}^{14}$ *is generated by 1 linear form, 6 quadrics, and 17 cubics.*

The equations, as well as the code we used to obtain them, can be found online at [Sey]. The following conjecture is closely related to [BM05, Conjecture 11.9] in algebraic statistics.

Conjecture 2.4.22. *For any fixed D and d, there is an M such that the ideal of* uMPS(D, d, N) *is generated by quadrics for all $N \geq M$.*

Using the trace parametrization, and the invariance of uMPS(D, d, N) under local transformations, we were able to obtain a complete description of the set uMPS$(2, 2, 4) \subseteq \mathrm{Cyc}^4(\mathbb{C}^2)$: it can be obtained by removing three GL_2-orbits from a degree six hypersurface in $\mathrm{Cyc}^4(\mathbb{C}^2) \cong \mathbb{C}^6$. For more details, see Section 2.4.5.

2.4.4 The fundamental theorem

In the literature appear several versions of the fundamental theorem of matrix product states, which all roughly say that for N large enough the map parametrizing matrix product states is generically injective up to obvious symmetry. The following formulation is adapted from [MGRPG$^+$18, Corollary 7].

Theorem 2.4.23. *Let $A_0, \dots, A_{d-1} \in \mathbb{C}^{D \times D}$ and $B_0, \dots, B_{d-1} \in \mathbb{C}^{D \times D}$ be such that $T_N(A_0, \dots, A_{d-1}) = T_N(B_0, \dots, B_{d-1})$ and assume that $N \geq 2L + 1$, where L is such that $\mathrm{span}(\{A_0, \dots, A_{d-1}\})^L = \mathrm{span}(\{B_0, \dots, B_{d-1}\})^L = \mathbb{C}^{D \times D}$. Then there is an invertible matrix Z and a constant $\zeta \in \mathbb{C}$ with $\zeta^N = 1$, such that $B_i = \zeta Z^{-1} A_i Z$ for every i. Moreover Z is unique up to a multiplicative constant.*

We recall the generic injectivity index $GC_{D,d}$ from Definition 2.3.8: it is the lowest number such that the following holds: for a generic tuple $\{A_0, \dots, A_{d-1}\}$ of $D \times D$-matrices, $\mathrm{span}(\{A_0, \dots, A_{d-1}\})^{GC_{D,d}} = \mathbb{C}^{D \times D}$.

Corollary 2.4.24. *Assume that $N \geq 2GC_{D,d} + 1$, then for a generic tensor in* uMPS(D, d, N) *all preimages under the map T_N are the same up to simultaneous conjugation and multiplication by an $N-$th root of unity.*

This would be a trivial corollary of Theorem 2.4.23 if we could restrict the map T_N to the dense open subset of tuples $\mathcal{A} \in (\mathbb{C}^{D \times D})^d$ satisfying $\mathrm{span}(\mathcal{A}^L) = \mathbb{C}^{D \times D}$. However, it is a priori not clear that the complement of this set cannot map to a dense subset of uMPS(D, d, N). In the rest of this section, we show that this indeed does not happen, using the trace parametrization.

Before we start the proof, we introduce the following notation: A point $\mathcal{A} = (A_0, \dots, A_{d-1}) \in (\mathbb{C}^{D \times D})^d$ gives rise to a D-dimensional representation $\varphi_{\mathcal{A}}$ of the associative algebra $\mathbb{C}\langle X_0, \dots, X_{d-1} \rangle$, by putting $\varphi_{\mathcal{A}}(X_i) = A_i$.

Theorem 2.4.25 ([KW85, II.2.7]).   • *The association $\mathcal{A} \mapsto \varphi_{\mathcal{A}}$ induces a bijection between the GL_D-orbits in $(\mathbb{C}^{D \times D})^d$, and the set of D-dimensional representations of $\mathbb{C}\langle X_0, \ldots, X_{d-1}\rangle$ up to isomorphism.*

    • *The orbit of $\mathcal{A}$ is closed if and only if $\varphi_{\mathcal{A}}$ is a semisimple representation.*

Corollary 2.4.26. *The orbit of a general point in $(\mathbb{C}^{D \times D})^d$ is closed.*

Proof. We claim that for a general point $\mathcal{A} = (A_0, \ldots, A_{d-1}) \in (\mathbb{C}^{D \times D})^d$, the representation $\varphi_{\mathcal{A}}$ is simple. Indeed: $\varphi_{\mathcal{A}}$ is simple if and only if there is no nontrivial subspace of $\mathbb{C}^D$ simultaneously fixed by all A_i, which is clearly the case for a generic choice of $\varphi_{\mathcal{A}}$. (For example: if all A_i are diagonalizable, having a subspace fixed by A_0 and A_1 would in particular imply that there is an eigenvector of A_1 that is a linear combination of less than D eigenvectors of A_0; a nongeneric condition.) $\qquad\square$

If $\pi : \operatorname{Spec} R \to \operatorname{Spec} R^G$ is a good categorical quotient, π is called a *geometric quotient* if every G-orbit in $\operatorname{Spec} R$ is closed, or equivalently, if π induces a bijection between orbits in $\operatorname{Spec} R$ and points in $\operatorname{Spec} R^G$. We call π is an *almost geometric quotient* if the orbit of a general point in $\operatorname{Spec} R$ is closed, or equivalently [CLS11, Prop 5.0.11.] if $\operatorname{Spec} R^G$ has a Zariski dense open subset U such that $\pi|_{\pi^{-1}(U)}$ is a geometric quotient.

Proof of Corollary 2.4.24. By the discussion above and Corollary 2.4.26, there is a dense open subset $U \subseteq \operatorname{Spec} \mathcal{C}_{D,d}$ such that $\pi|_{\pi^{-1}(U)}$ induces a bijection between GL_D-orbits in $\pi^{-1}(U)$ and points in U.

We show that $\operatorname{Spec} \mathcal{C}_{D,d}$ and $\mathrm{uMPS}(D, d, N)$ both have dimension $(d-1)D^2+1$. For $\operatorname{Spec} \mathcal{C}_{D,d}$ the dimension can be computed as the difference $\dim((\mathbb{C}^{D \times D})^d) - \dim(\pi^{-1}(x))$, where x is a general point in $\operatorname{Spec} \mathcal{C}_{D,d}$. We can assume $x \in U$, so that $\pi^{-1}(x)$ is the GL_D-orbit of a generic d-tuple of matrices, which clearly has dimension $D^2 - 1$. For $\mathrm{uMPS}(D, d, N)$: let $V \subseteq (\mathbb{C}^{D \times D})^d$ be the set of tuples $\mathcal{A}$ for which $\operatorname{span}(\mathcal{A})^{GC_{D,d}} = \mathbb{C}^{D \times D}$. By the definition of $GC_{D,d}$, V is dense, hence $\dim \mathrm{uMPS}(D, d, N) = \dim(T_N(V))$. Now we can do the same computation as before: by Theorem 2.4.23, a general fiber of $T_N|_V$ has dimension $D^2 - 1$, and we conclude $\dim(\mathrm{uMPS}(D, d, N)) = (d - 1)D^2 + 1$.

We claim that $\pi((\mathbb{C}^{D \times D})^d \setminus V)$ is contained in a lower-dimensional subspace of $\operatorname{Spec} \mathcal{C}_{D,d}$. Indeed, consider $Y = \pi((\mathbb{C}^{D \times D})^d \setminus V) \cap U$. Then since V is GL_D-invariant and by definition of U, it holds that $\pi^{-1}(Y) = ((\mathbb{C}^{D \times D})^d \setminus V) \cap \pi^{-1}(U)$. So $\pi^{-1}(Y)$ is not Zariski dense, hence the same holds for $\pi((\mathbb{C}^{D \times D})^d \setminus V)$.

Now consider the map $\widetilde{T_N} : \operatorname{Spec} \mathcal{C}_{D,d} \twoheadrightarrow \mathrm{uMPS}(D, d, N)$. Since both spaces have the same dimension, the lower-dimensional set $\pi((\mathbb{C}^{D \times D})^d \setminus V)$ will map to a lower-dimensional subset of $\mathrm{uMPS}(D, d, N)$. So for a general point $x \in \mathrm{uMPS}(D, d, N)$, we get $T_N^{-1}(x) \subseteq V$. Then we are done by Theorem 2.4.23. $\qquad\square$

The above Corollary can also be stated as follows: assume that $N \geq 2GC_{D,d}+1$, then for a generic tensor in uMPS(D, d, N) there are exactly N preimages under $\widetilde{T_N}$. In the case $D = d = 2$, one easily checks that $GC_{D,d} = 2$, yielding the following result, which was stated as a conjecture in [CM14, Conjecture 12].

Theorem 2.4.27. *Using the trace parametrization φ_N, for $N \geq 5$ almost every matrix product state in* uMPS$(2, 2, N)$ *has exactly N choices of parameters that yield it.*

2.4.5 uMPS$(2, 2, 4)$ as a constructible set

In this section we give a description of uMPS$(2, 2, 4)$ as a constructible subset of $\mathrm{Cyc}^4(\mathbb{C}^2) \cong \mathbb{C}^6$. See [Sey] for Macaulay2 code accompanying this section. Using the notation from Section 2.2, $\mathrm{Cyc}^4(\mathbb{C}^2)$ has a basis given by e_{0000}, e_{0001}, e_{0011}, e_{0111}, e_{1111}, $\underline{e_{0101}}$. We write coordinates on the space $\mathrm{Cyc}^4(\mathbb{C}^2)$ by $x_{0000}, \ldots, x_{0101}$. The closure $\overline{\mathrm{uMPS}(2, 2, 4)}$ was already computed in [CM14, Theorem 3]: it is a hypersurface cut out by the polynomial

$$
\begin{aligned}
f_{224} = \ &2x_{0011}^6 - 12x_{0001}x_{0011}^4x_{0111} + 16x_{0001}^2x_{0011}^2x_{0111}^2 + 4x_{0000}x_{0011}^3x_{0111}^2 \\
&- 8x_{0000}x_{0001}x_{0011}x_{0111}^3 + x_{0000}^2x_{0111}^4 + 4x_{0001}^2x_{0011}^3x_{1111} - x_{0000}x_{0011}^4x_{1111} \\
&- 8x_{0001}^3x_{0011}x_{0111}x_{1111} + 2x_{0000}x_{0001}^2x_{0111}^2x_{1111} + x_{0001}^4x_{1111}^2 \\
&+ 8x_{0001}x_{0011}^3x_{0111}x_{0101} - 16x_{0001}^2x_{0011}x_{0111}^2x_{0101} - 4x_{0000}x_{0011}^2x_{0111}^2x_{0101} \\
&+ 4x_{0000}x_{0001}x_{0111}^3x_{0101} - 4x_{0001}^2x_{0011}^2x_{1111}x_{0101} + 4x_{0001}^3x_{0111}x_{1111}x_{0101} \\
&+ 8x_{0000}x_{0001}x_{0011}x_{0111}x_{1111}x_{0101} - 2x_{0000}^2x_{0111}^2x_{1111}x_{0101} \\
&- 2x_{0000}x_{0001}^2x_{1111}^2x_{0101} - x_{0011}^4x_{0101}^2 + 4x_{0001}^2x_{0111}^2x_{0101}^2 + 4x_{0000}x_{0011}x_{0111}^2x_{0101}^2 \\
&+ 4x_{0001}^2x_{0011}x_{1111}x_{0101}^2 - 2x_{0000}x_{0011}^2x_{1111}x_{0101}^2 - 4x_{0000}x_{0001}x_{0111}x_{1111}x_{0101}^2 \\
&+ x_{0000}^2x_{1111}^2x_{0101}^2 - 2x_{0000}x_{0111}^2x_{0101}^3 - 2x_{0001}^2x_{1111}x_{0101}^3 + x_{0000}x_{1111}x_{0101}^4
\end{aligned}
$$

as can be verified by a Gröbner basis computation, for example in Macaulay2. In principle, we could use *TotalImage.m2* (see [HMS19]) to compute the image of the trace parametrization map $\widetilde{T} : \mathbb{C}^5 \to \mathrm{Cyc}^4(\mathbb{C}^2)$. This computation did not finish in a reasonable amount of time. However, as we will explain now, one can exploit symmetries of uMPS$(2, 2, 4)$ to simplify the computations.

Recall from Remark 2.2.14 that uMPS$(2, 2, 4)$ is invariant under the natural GL_2-action on $\mathrm{Cyc}^4(\mathbb{C}^2)$. We use the following strategy

1. Find a low-dimensional subset $Y \subseteq \mathrm{Cyc}^4(\mathbb{C}^2)$ that contains at least one point from every GL_2-orbit.

2. Use *TotalImage.m2* to compute $Z = \widetilde{T}(\widetilde{T}^{-1}(Y))$.

3. Compute $GL_2 \cdot (Y \cap Z)$.

We now describe this in more detail. First, note that

$$\mathrm{Cyc}^4(\mathbb{C}^2) \cong \mathrm{Sym}^4(\mathbb{C}^2) \oplus \mathbb{C} \tag{2.4.4}$$

where the map from $\mathbb{C}$ to $\mathrm{Cyc}^4(\mathbb{C}^2)$ is given by sending 1 to $e_{0011} - 2e_{0101}$ and the map from $\mathrm{Sym}^4(\mathbb{C}^2)$ to $\mathrm{Cyc}^4(\mathbb{C}^2)$ is given by

$$x^4 \mapsto e_{0000}$$
$$x^3 y \mapsto \frac{1}{4} e_{0001}$$
$$x^2 y^2 \mapsto \frac{1}{6}(e_{0011} + e_{0101})$$
$$xy^3 \mapsto \frac{1}{4} e_{0111}$$
$$y^4 \mapsto e_{1111}.$$

Because $\mathrm{Sym}^4(\mathbb{C}^2)$ can be seen as the space of homogeneous degree 4 polynomials in 2 variables one can easily see that the following set contains exactly one representative of every GL_2−orbit

$$\{x^4, x^3 y, x^2 y^2\} \cup \{xy(x - y)(x - \mu y) \mid \mu \in \mathbb{C} \setminus \{1\}\}.$$

Using the isomorphism (2.4.4), one can deduce that if we define the following subsets of $\mathrm{Cyc}^4(\mathbb{C}^2)$

$$Y_1 = V(x_{0001}, x_{0111}, x_{1111}, 2x_{0011} + x_{0101})$$
$$Y_2 = V(x_{0000}, x_{0111}, x_{1111}, 2x_{0011} + x_{0101})$$
$$Y_3 = V(x_{0000}, x_{0001}, x_{0111}, x_{1111})$$
$$Y_4 = V(x_{0000}, x_{1111}, 2x_{0001} + 2x_{0011} + 2x_{0111} + x_{0101})$$

then $Y = Y_1 \cup Y_2 \cup Y_3 \cup Y_4$ contains at least one point from every GL_2-orbit. Using *TotalImage.m2*, we computed $Z_i := \widetilde{T}(\widetilde{T}^{-1}(Y_i))$ and compared it with $V(f_{224}) \cap Y_i$:

$$Z_1 = \{e_{0000}\} = V(f_{224}) \cap Y_1$$
$$Z_2 = \emptyset \neq \{e_{0001}\} = V(f_{224}) \cap Y_2$$
$$Z_3 = \{e_{0101}\} \neq \{e_{0101}, e_{0011} + \sqrt{2}e_{0101}, e_{0011} - \sqrt{2}e_{0101}\} = V(f_{224}) \cap Y_3$$
$$Z_4 = V(f_{224}) \cap Y_4.$$

We conclude the following: uMPS$(2, 2, 4)$ is the vanishing locus of the polynomial f_{224}, with the orbits of the following three tensors removed: e_{0001}, $e_{0011} + \sqrt{2}e_{0101}$, $e_{0011} - \sqrt{2}e_{0101}$. One can easily compute the orbit closures: they are cut out by the following ideals:

$$I_1 = (x_{0011} - x_{0101},\, x_{0000}x_{1111} - 4x_{0001}x_{0111} + 3x_{0101}^2,$$
$$x_{0000}x_{0111}^2 + x_{0001}^2 x_{1111} - 6x_{0001}x_{0111}x_{0101} + 4x_{0101}^3,$$
$$x_{0001}^2 x_{1111}^2 + 4x_{0001}x_{0111}^3 - 6x_{0001}x_{0111}x_{1111}x_{0101} - 3x_{0111}^2 x_{0101}^2 + 4x_{1111}x_{0101}^3)$$

$$I_2 = (x_{0011}x_{1111} + \sqrt{2}x_{0111}^2 + (-1 - \sqrt{2})x_{1111}x_{0101},$$
$$x_{0001}x_{1111} + (-2 - \sqrt{2})x_{0011}x_{0111} + (1 + \sqrt{2})x_{0111}x_{0101},$$
$$2x_{0001}x_{0111} + (2 + 2\sqrt{2})x_{0011}^2 + (-8 - 5\sqrt{2})x_{0011}x_{0101} + (4 + 3\sqrt{2})x_{0101}^2,$$
$$x_{0000}x_{1111} + (-6 - 4\sqrt{2})x_{0011}^2 + (8 + 6\sqrt{2})x_{0011}x_{0101} + (-3 - 2\sqrt{2})x_{0101}^2,$$
$$x_{0000}x_{0111} + (-2 - \sqrt{2})x_{0001}x_{0011} + (1 + \sqrt{2})x_{0001}x_{0101},$$
$$x_{0000}x_{0011} + (-1 - \sqrt{2})x_{0000}x_{0101} + \sqrt{2}x_{0001}^2)$$

$$I_3 = (x_{0011}x_{1111} - \sqrt{2}x_{0111}^2 + (-1 + \sqrt{2})x_{1111}x_{0101},$$
$$x_{0001}x_{1111} + (-2 + \sqrt{2})x_{0011}x_{0111} + (1 - \sqrt{2})x_{0111}x_{0101},$$
$$2x_{0001}x_{0111} + (2 - 2\sqrt{2})x_{0011}^2 + (-8 + 5\sqrt{2})x_{0011}x_{0101} + (4 - 3\sqrt{2})x_{0101}^2,$$
$$x_{0000}x_{1111} + (-6 + 4\sqrt{2})x_{0011}^2 + (8 - 6\sqrt{2})x_{0011}x_{0101} + (-3 + 2\sqrt{2})x_{0101}^2,$$
$$x_{0000}x_{0111} + (-2 + \sqrt{2})x_{0001}x_{0011} + (1 - \sqrt{2})x_{0001}x_{0101},$$
$$x_{0000}x_{0011} + (-1 + \sqrt{2})x_{0000}x_{0101} - \sqrt{2}x_{0001}^2)$$

Finally, one can check that $V(I_1) = (GL_2 \cdot e_{0001}) \cup (GL_2 \cdot e_{0000})$, $V(I_2) = (GL_2 \cdot (e_{0011} + \sqrt{2}e_{0101})) \cup (GL_2 \cdot e_{0000})$, and $V(I_3) = (GL_2 \cdot (e_{0011} - \sqrt{2}e_{0101})) \cup (GL_2 \cdot e_{0000})$. Now $GL_2 \cdot e_{0000}$, is a closed orbit consisting of all rank 1 symmetric tensors in $\mathrm{Cyc}^4(\mathbb{C}^2)$. Explicitly, it is cut out by the ideal

$$J = (x_{0101} - x_{0011}, x_{0000}x_{0011} - x_{0001}^2, x_{0000}x_{0111} - x_{0001}x_{0011}, x_{0000}x_{1111}$$
$$- x_{0001}x_{0111}, x_{0001}x_{0111} - x_{0011}^2, x_{0001}x_{1111} - x_{0011}x_{0111}, x_{0011}x_{1111} - x_{0111}^2).$$

Finally, we obtain the following description of $\mathrm{uMPS}(2, 2, 4) \subseteq \mathrm{Cyc}^4(\mathbb{C}^2)$ as a constructible set:

$$\mathrm{uMPS}(2, 2, 4) = (V(f_{224}) \setminus (V(I_1) \cup V(I_2) \cup V(I_3))) \cup V(J).$$

Remark 2.4.28. Very recently, in [BLH19], Barakat and Lange-Hegermann computed a description of $\mathrm{uMPS}(2, 2, 5)$ as a constructible set. Their computational methods also allow to compute the description of $\mathrm{uMPS}(2, 2, 4)$ above without needing to use the additional GL_2-symmetry.

2.5 A tensor version of The Quantum Wielandt theorem

Projected entangled pair states (PEPS) were defined in Definition 2.1.7. They are higher-dimensional generalizations of matrix product states, and play a central role in the classification of the different quantum phases of spin systems defined on two-dimensional grids. PEPS are much more complex than MPS: just as MPS can be understood in terms of completely positive maps on matrices, PEPS deal with completely positive maps on tensors, for which no analogues of eigenvalue and singular value decompositions exist. It has been a long standing open question in the field of quantum tensor networks whether an analogue of the quantum Wielandt theorem exists for PEPS, which is the missing piece in proving that every PEPS has a parent Hamiltonian with finite support — cf. [CGRPG19, Section 2] and references therein. In this section, we prove the existence of such a theorem, albeit in a weaker form than for MPS as the upper bound is nonconstructive. In physics terms, it is proven that the notion of injectivity for PEPS is well defined, in the sense that there is only a finite amount of blocking needed for the map from the virtual to the physical indices to become injective. Our proof is a natural generalization of the proof in Section 2.3 for the existence of an injectivity radius (see Lemma 2.3.1, Lemma 2.3.6, and Theorem 2.3.7).

We fix natural numbers n (grid dimension), D (bond dimension), d (physical dimension). The case $n = 1$ will correspond to the setup in Section 2.3. In this section, we will fix a space $V := (\mathbb{C}^D)^{\otimes 2n} \otimes \mathbb{C}^d$. Tensors in V will be identified with d-tuples of tensors in $(\mathbb{C}^D)^{\otimes 2n}$. For $A \in V$, its slices will be denoted by $A_1, \ldots, A_d \in (\mathbb{C}^D)^{\otimes 2n}$; the linear span of these slices will be denoted by $\langle A \rangle \subseteq (\mathbb{C}^D)^{\otimes 2n}$. We want to define a space $\mathcal{A}^{(N_1,\ldots,N_n)}$, as the linear span of all tensors obtained by arranging a collection of tensors in $\mathcal{A}$ in a grid of size $N_1 \times \cdots \times N_n$, and contracting over the edges. In the following paragraph, we will make this more precise.

For $N_1, \ldots, N_n \in \mathbb{N}$, we define the graph $\Gamma = \Gamma(N_1, \ldots, N_n)$ to be the n-dimensional square grid of size $N_1 \times \ldots \times N_n$, with an aditional outgoing edge at every vertex. These additional edges will be called *physical edges*, the other edges will be called *virtual edges*. The grid $\Gamma(3, 5)$ is presented below:

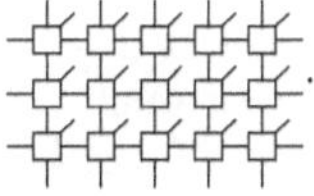

We will denote the outgoing virtual edges of Γ by $(\boldsymbol{j}, \pm e_i)$, where $\boldsymbol{j}$ is a vertex on the boundary of the grid, and $\pm e_i$ indicates the direction of the outgoing edge. The set of outgoing virtual edges of G will be denoted by $E_O(\Gamma)$.

To every virtual edge e we associate the same vector space $V_e = \mathbb{C}^D$, and to every physical edge e we associate the same vector space $V_e = \mathbb{C}^D$. Now we can identify all virtual spaces $V_v = (\mathbb{C}^D)^{\otimes 2n} \otimes (\mathbb{C}^d)$ in the obvious way: the tensor factor associated to an edge out of v will be identified with the tensor factor associated to the edge out of w pointing in the same direction. We also fix an identification of all these spaces V_v with our fixed space V. Having done all these identifications, we can now define the space $\mathcal{A}^{(N_1,\ldots,N_n)}$ as follows: place the tensor $\mathcal{A}$ at every vertex of $\Gamma(N_1, \ldots, N_n)$, and contract over the inner edges to obtain a tensor T in $(\mathbb{C}^D)^{\otimes |E_O(\Gamma)|} \otimes (\mathbb{C}^d)^{\otimes N}$ (where $N := \prod N_i$). Then $\mathcal{A}^{(N_1,\ldots,N_n)} \subseteq (\mathbb{C}^D)^{\otimes |E_O(\Gamma)|}$ is defined to be the linear span of the slices of T. In what follows, we will use the notation $\mathcal{C}[(v \mapsto B_v)_{v \in \Gamma}]$, to denote the contraction of a collection of tensors $B_v \in (\mathbb{C}^D)^{\otimes 2n}$ placed at the vertices of Γ.

Definition 2.5.1. We say that $\Gamma(N_1, \ldots, N_n)$ is a *spanning region* for $\mathcal{A}$ if $\mathcal{A}^{(N_1,\ldots,N_n)} = (\mathbb{C}^D)^{\otimes |E_O(\Gamma)|}$. Equivalently, Γ is a spanning region for $\mathcal{A}$, if the tensors $\mathcal{C}[(v \mapsto A_{i_v})_{v \in \Gamma}]$ linearly span the whole space $(\mathbb{C}^D)^{\otimes E_O(\Gamma)}$, when we consider all possible ways of placing a tensor from $\mathcal{A}$ at every vertex of G. If $\mathcal{A}$ has an spanning region, we say that $\mathcal{A}$ is *spanning*.

Remark 2.5.2. In the case $n = 1$, the definition of the space $\mathcal{A}^N$ agrees with the space A^N defined in section 2.3. Moreover, $\Gamma(N)$ is an spanning region for $\mathcal{A}$ if and only if N is greater than or equal to the spanning index $i(\mathcal{A})$.

Remark 2.5.3. We note that being an spanning region for G and being spanning are properties of the linear span of $\mathcal{A}$, not a particular choice of tensors A_i.

In the following lemma, which is a generalization of Lemma 2.3.1, we prove that being an spanning region is stable under extension of the grid.

Lemma 2.5.4. *Let $G_1 \subseteq G_2$ be n-dimensional square grids. If G_1 is an spanning region for $\mathcal{A}$, then so is G_2.*

Proof. By induction, we may assume that $G_1 = G(N_1 - 1, N_2, \ldots, N_n)$ and $G_2 = G(N_1, N_2, \ldots, N_n)$. If $N_1 = 2$ the statement is true, because G_2 is the union of two spanning regions, cf [PGVWC08, Lemma 1]. Thus we assume $N_1 > 2$. The vertices of G_2 will be identified with vectors $\boldsymbol{j} = (j_1, \ldots, j_n) \in \mathbb{N}^n$, with $1 \leq j_i \leq N_i$. Such a vertex is in G_1 if additionally $j_1 \leq N_1 - 1$. We need to show that every tensor $T \in V^{\otimes E_O(G_2)}$ can be written as a linear combination of tensors of the form $\mathcal{C}[(\boldsymbol{j} \mapsto A_{i_{\boldsymbol{j}}})_{\boldsymbol{j} \in G_2}]$. In fact it is enough to show this for rank one tensors T, since every tensor is a linear combination of rank one tensors.

We can identify $E_O(G_1)$ with a subset of $E_O(G_2)$ as follows: to an outgoing edge $(\boldsymbol{j}, \pm e_i)$ of $E_O(G_1)$, we associate $(\boldsymbol{j}, \pm e_i)$ if $\pm e_i \neq +e_1$, and $(\boldsymbol{j} + e_1, +e_1)$ if $\pm e_i = +e_1$. Assuming T has rank one, we have $T = T_1 \otimes T_2 \in V^{\otimes E_O(G_1)} \otimes V^{\otimes r}$, where r equals the cardinality of $E_O(G(N_2, \ldots, N_n))$.

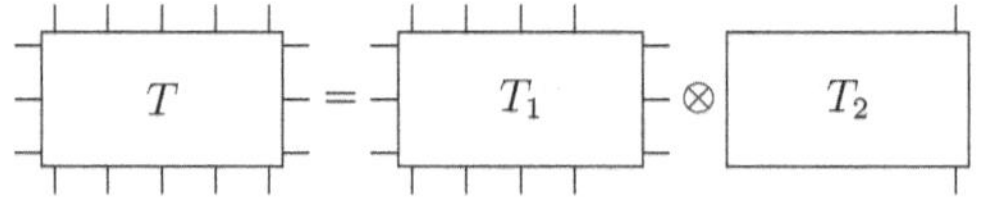

By assumption we can write T_1 as a linear combination of tensors of the form $\mathcal{C}[(\boldsymbol{j} \mapsto A_{i_{\boldsymbol{j}}})_{\boldsymbol{j} \in G_1}]$. Let G_1' be the grid obtained from G_1 by contracting all inner edges among vertices $\boldsymbol{j}$ for which $j_1 > 1$. This grid is the rightmost one in the picture below. In particular, all vertices with $j_1 > 1$ get identified to a vertex v_1. Then T_1 is in particular a linear combination of tensors of the form

$\mathcal{C}[(v \mapsto B_v)_{v \in G_1'}]$, where $B_v = A_{i_v}$ if v is one of the vertices that did not get contracted and $B_{v_1} = \mathcal{C}[(\boldsymbol{j} \mapsto A_{i_{\boldsymbol{j}}})_{\boldsymbol{j} \in G_1, j_1 > 1}]$.

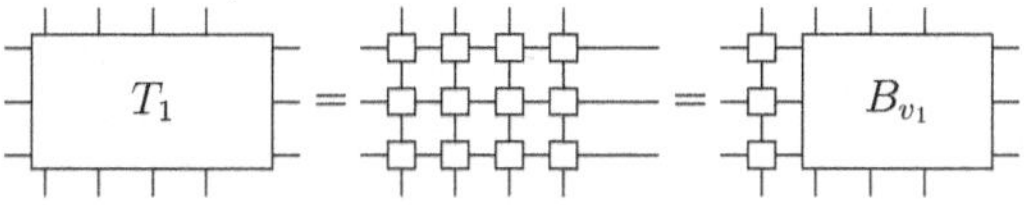

Consider the tensors $B_{v_1} \otimes T_2 \in V^{\otimes |E_O(G_1)|}$. By assumption each one is a linear combination of tensors of the form $\mathcal{C}[(\boldsymbol{j} \mapsto A_{k_{\boldsymbol{j}}})_{\boldsymbol{j} \in G_1}]$, where now we identified G_1 with the subgrid of G_2 consisting of all vertices $\boldsymbol{j}$ with $j_1 > 1$.

Thus, we see that T is a combination $\mathcal{C}[(\boldsymbol{j} \mapsto A_{s_{\boldsymbol{j}}})_{\boldsymbol{j} \in G_2}]$ where s may be identified with i above for $\boldsymbol{j}$ such that $j_1 = 1$ and with k when $j_1 > 1$.

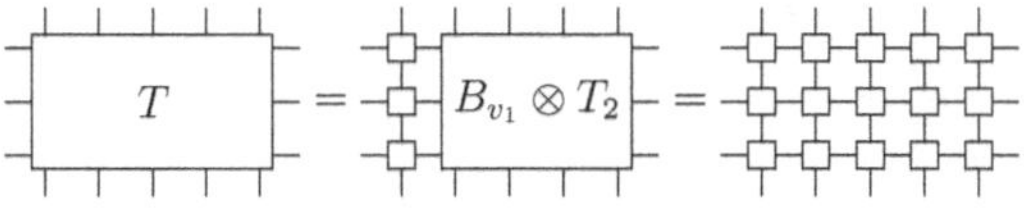

Our main theorem says that if $\mathcal{A}$ is spanning, then there exists an spanning region of bounded size (where the bound only depends on our parameters D, d, n).

Theorem 2.5.5. *Let $G_1 \subset G_2 \subset \cdots \subset G_k \subset \cdots$ be a chain of n-dimensional grids. Then there exists a constant C (depending on D, d, and the chain) such that the following holds:*

If $\mathcal{A} \in (\mathbb{C}^D)^{\otimes 2n} \otimes \mathbb{C}^d$ is chosen so that for some k, G_k is an spanning region for $\mathcal{A}$, then already G_C is an spanning region for $\mathcal{A}$.

Proof. For any grid G and $\mathcal{A} \in ((\mathbb{C}^D)^{\otimes 2n})^d$, we write

$$S_G(\mathcal{A}) := \{\mathcal{C}[(v \mapsto A_{i_v})_{v \in G}]\} \subseteq (\mathbb{C}^D)^{\otimes E_O(\Gamma)},$$

and

$$V_G := \{\mathcal{A} \in ((\mathbb{C}^D)^{\otimes 2n})^d \,|\, \text{span}(S_G(\mathcal{A})) \subsetneq (\mathbb{C}^D)^{\otimes E_O(\Gamma)}\}.$$

Thus, G is an spanning region for $\mathcal{A}$ if and only if $\text{span}(S_G(\mathcal{A})) = (\mathbb{C}^D)^{\otimes E_O(\Gamma)}$ if and only if $\mathcal{A} \notin V_G$. By Lemma 2.5.4, it holds that $V_{G_1} \supseteq V_{G_2} \supseteq \cdots \supseteq V_{G_k} \supseteq \cdots$ We need to show that this chain eventually stabilizes. We will show that every V_{G_k} is a Zariski closed subset of $((\mathbb{C}^D)^{\otimes 2n})^d$, i.e. that is the zero locus of a system of polynomials. This will finish the proof by Hilbert Basis Theorem.

Fix a grid $G = G_k$. For every $\mathcal{A} \in ((\mathbb{C}^D)^{\otimes 2n})^d$, we can build a $D^{E_O(\Gamma)} \times d^{|\mathcal{V}(G)|}$ matrix $M_{\mathcal{A}}$ whose entries are the coefficients of the elements of $S_G(\mathcal{A})$. The condition $\text{span}(S_G(\mathcal{A})) \subsetneq (\mathbb{C}^D)^{\otimes E_O(\Gamma)}$ is equivalent to $M_{\mathcal{A}}$ having rank smaller than $D^{E_O(\Gamma)}$. The entries of $M_{\mathcal{A}}$ are polynomials in the entries of $\mathcal{A}$. Hence, the condition $\mathcal{A} \in V_G$ can be expressed as the vanishing of certain polynomials ($D^{E_O(\Gamma)}$-minors of $M_{\mathcal{A}}$) in the entries of $\mathcal{A}$. Hence, V_G is a Zariski closed subset of $((\mathbb{C}^D)^{\otimes 2n})^d$. $\qquad\square$

Theorem 2.5.5 can be reformulated as follows.

Theorem 2.5.6. *There exists a finite collection of grids $G_1, \ldots, G_M$ (depending on n, D, d) such that the following holds:*

> *If $\mathcal{A} \in ((\mathbb{C}^D)^{\otimes 2n})^d$ is spanning, then one of the G_i is an spanning region for $\mathcal{A}$.*

The equivalence of Theorem 2.5.5 and Theorem 2.5.6 follows from the following general lemma.

Lemma 2.5.7. *Let $\mathcal{P}$ be a partially ordered set. We consider $\mathbb{N}^n$ with the coordinatewise partial order. Let $f : \mathbb{N}^n \to \mathcal{P}$ be a map such that*

1. *$a_1 \leq a_2 \implies f(a_1) \geq f(a_2)$.*

2. *For every chain $a_1 < a_2 < \ldots$ in $\mathbb{N}^n$, the chain $f(a_1) \geq f(a_2) \geq \ldots$ stabilizes after finitely many steps.*

Then there is a finite subset B of $\mathbb{N}^n$ such that for any $a \in \mathbb{N}^n$, there is a $b \in B$ with $a \geq b$ and $f(a) = f(b)$.

Proof. We first claim that there is a $b_0 \in \mathbb{N}^n$ such that $f(a) = f(b_0)$ for every $a \geq b_0$. Indeed, if there was no such b_0 we could build an infinite chain $a_1 < a_2 < \ldots$ in $\mathbb{N}^n$ with $f(a_1) > f(a_2) > \ldots$

Now we can proceed by induction on n: the subset $\{a \in \mathbb{N}^n | a \not\geq b_0\}$ can be written as a finite union of hyperplanes, each of which can be identified with $\mathbb{N}^{n-1}$. By the induction hypothesis, in each such hyperplane $H \subset \mathbb{N}^n$ there is a finite subset $B_H \subset H$ such that for any $a \in H$, there is a $b \in B_H$ with $a \geq b$ and $f(a) = f(b)$. We define B as b_0 together with the union of all B_H. $\square$

Proof of Theorem 2.5.6. We apply Lemma 2.5.7 by identifying $\mathbb{N}^n$ with the collection of n-dimensional grids and taking $\mathcal{P}$ to be the poset of subsets of $((\mathbb{C}^D)^{\otimes 2n})^d$ ordered by inclusion, and $f : G \mapsto V_G$, where V_G was defined in the proof of Theorem 2.5.5. We conclude by Theorem 2.5.5. $\square$

We note that the constants in Theorem 2.5.5 and 2.5.6 can be chosen independent of d.

Corollary 2.5.8. *For any n, D there exists a finite collection of grids $G_1, \ldots, G_M$ such that the following holds:*

> *For any d, if $\mathcal{A} \in ((\mathbb{C}^D)^{\otimes 2n})^d$ is spanning, then one of the G_i is an spanning region for $\mathcal{A}$.*

Proof. By Remark 2.5.3 it is enough to consider the subspaces $\langle \mathcal{A} \rangle \subset ((\mathbb{C}^D)^{\otimes 2n})$. In particular the dimension of the subspaces is bounded by D^{2n} and for each fixed dimension we obtain a finite number of grids by Theorem 2.5.6. $\square$

Further we have the following computational implication.

Corollary 2.5.9. *For every fixed n and D, there exists an algorithm to decide if $\mathcal{A}$ is spanning.*

Proof. Let $G_1, \ldots, G_M$ be the set of grids from Corollary 2.5.8. The algorithm checks for every i whether G_i is an spanning region for $\mathcal{A}$. For a fixed grid, this amounts to checking surjectivity of a given polynomial map, cf. [HMS19]. We know $\mathcal{A}$ is spanning if and only if it is spanning for one of the G_i. $\square$

We note that although we know such an algorithm exists, we cannot explicitly provide it. The reason is that we do not know the grids $G_1, \ldots, G_M$ from Corollary 2.5.8 - we just know they exist. Our result should also be contrasted with [SMG+18, Theorem 4], which states that there is no algorithm that receives ϕ_v and decides if $\mathcal{C}[(v \mapsto \phi_v)_{v \in T_{x,y}}] = 0 \ \forall x, y \in \mathbb{N}$, where $T_{x,y}$ is the $x \times y$-torus.

2.6 Future directions

In Section 2.5, there is an obvious question left open.

Question 2.6.1. *Can Theorem 2.5.6 be made constructive?*

More precisely, it follows from Theorem 2.5.6 that for every n, D, d, there exists a constant $C := C_{n,D,d}$ such that if $\mathcal{A} \in ((\mathbb{C}^D)^{\otimes 2n})^d$ is spanning, then a grid of size $C \times \cdots \times C$ is a spanning region for $\mathcal{A}$. For $n = 1$, this constant is the injectivity radius, which we know to be $O(D^2 \log D)$. But even for $n = 2$, we have no complexity bound whatsoever. Regarding the defining equations of uMPS$(2, 2, N)$, we have the following conjecture.

Conjecture 2.6.2. *There is a constant c, so that for $N \gg 0$, the ideal defining* uMPS$(2, 2, N)$ *is generated in degree $\leq c$.*

In fact, we suspect that for $N \geq 6$, the ideal of uMPS$(2, 2, N)$ is generated in degree ≤ 3. By Theorem 2.4.21, this is true for $N = 6$, and preliminary computations indicate it holds for $N = 7$ as well. Conjecture 2.6.2 is inspired by [BM05, Conjecture 11.9], which states that the ideal defining so-called *binary hidden Markov models* is generated in degree 2. It would be interesting to study the connections between (uniform) matrix product states and hidden Markov models more closely.

By Remark 2.2.14, the degree r part of the ideal of uMPS(D, d, N) is a representation of GL_d for every $r \in \mathbb{N}$. One plan for a future project is to study these representations more closely, in the hope that they reveal more information about the structure of uMPS(D, d, N). In particular, this might be helpful for finding a proof of (or a counterexample to) Conjecture 2.6.2.

Chapter 3

Fast Matrix Multiplication

In this chapter, we present several contributions to the study of fast matrix multiplication. The complexity of matrix multiplication is measured by the constant ω, which is defined as the smallest number such that for any $\epsilon > 0$ the multiplication of $n \times n$ matrices can be performed in time $O(n^{\omega+\epsilon})$. Trivially, it holds that $2 \leq \omega \leq 3$: the first inequality holds because the input size of matrix multiplication is already $\Theta(n^2)$, and the second inequality is given by the standard high school algorithm for matrix multiplication, which has complexity $O(n^3)$. Matrix multiplication can be viewed as a 3-way tensor –the *matrix multiplication tensor*– and estimating ω is equivalent to estimating the (border) rank of this tensor.

The best known upper bounds on ω were proven using the so-called *laser method*, originally due to Strassen and later refined by Coppersmith and Winograd. The main idea of the laser method is to, instead of directly studying the matrix multiplication tensor, consider a different tensor T which has both has low border rank and contains (in a sense to be made precise later) many copies of matrix multiplication tensors. In Section 3.1, we review the laser method, and prove a slight variant from the version usually found in the literature.

For studying fast matrix multiplication, one can instead of considering the usual matrix multiplication tensor, consider the *symmetrized matrix multiplication tensor*, which is a polynomial in $S^k(\mathfrak{sl}_n)$. Motivated by this, we study in Section 3.2 the plethysm $S^k(\mathfrak{sl}_n)$ of the adjoint representation $\mathfrak{sl}_n$ of the Lie group SL_n. In particular, we describe the decomposition of this representation into irreducible components for $k = 3$, and find highest weight vectors for all irreducible components. A surprising observation is that several of the highest weight vectors are, when viewed as tensors, equal to the *Coppersmith-Winograd tensor*, the best known tensor for applying the laser method to.

Recently, it was proven that it is not possible to prove $\omega < 2.3$ by applying the laser method to the Coppersmith-Winograd tensor [AFLG15]. This motivates a search for new, related tensors that are suitable for the laser method. In Section 3.3, we explore two ways of constructing such tensors. The first one builds upon the work of Landsberg–Michałek [LM17] and Bläser–Lysikov [BL16].

The idea is to obtain tensors of minimal border rank from the multiplication maps of finite-dimensional algebras. The second way is motivated by our observations in Section 3.2: since the CW-tensor arises as one of the highest weight vectors we obtained, it is natural to ask if some of the other highest weight vectors give rise to interesting tensors. We will focus on one of them in particular which appears to be well-suited for the laser method.

The results in Section 3.1 were all known before, though our formulation of Theorem 3.1.20 is new. Section 3.2 is based on the paper [Sey18], and Section 3.3 is based on joint work in progess with Joachim Jelisiejew and Mateusz Michałek.

3.1 The laser method

This section is an introduction to fast matrix multiplication and the laser method. Our exposition is based on [BCS97, Chapter 15] and [Lan17, Chapter 3].

Multiplication of matrices is a bilinear map $\mathbb{C}^{a\times b} \times \mathbb{C}^{b\times c} \to \mathbb{C}^{a\times c}$, or equivalently, a tensor in $\mathbb{C}^{a\times b} \otimes \mathbb{C}^{b\times c} \otimes \mathbb{C}^{c\times a}$. This tensor is known as the *matrix multiplication tensor* and denoted by $M_{\langle a,b,c\rangle}$. In coordinates:

$$M_{\langle a,b,c\rangle} = \sum_{i,j,k} e_{i,j} \otimes e_{j,k} \otimes e_{k,i}.$$

We will abbreviate $M_{\langle n,n,n\rangle}$ to $M_{\langle n\rangle}$. A rank r decomposition for $M_{\langle a,b,c\rangle}$ gives rise to an algorithm for matrix multiplication using r scalar multiplications. For example, Strassen's algorithm [Str69] for multiplying two 2×2 matrices using 7 scalar multiplications corresponds to a rank 7 decomposition of $M_{\langle 2,2,2\rangle}$. The following theorem was already known by Strassen, see also [BCS97, (15.1)].

Theorem 3.1.1. *The matrix multiplication constant ω satisfies*

$$\omega = \inf\{\tau \in \mathbb{R} \mid \mathrm{rk}(M_{\langle n\rangle}) = O(n^\tau)\}.$$

In fact, as shown by Bini, one can replace rank by border rank.

Theorem 3.1.2 ([Bin80], see also [Lan17, Theorem 3.2.1.10]).

$$\omega = \inf\{\tau \in \mathbb{R} \mid \underline{\mathrm{rk}}(M_{\langle n\rangle}) = O(n^\tau)\}.$$

3.1.1 Asymptotic sum inequality and degeneracy value

An important result used to obtain upper bounds on ω is Schönhage's *asymptotic sum inequality*.

Theorem 3.1.3 ([Sch81], see also [BCS97, (15.11)]).

$$\sum_i (a_i b_i c_i)^{\omega/3} \leq \underline{\mathrm{rk}}\left(\bigoplus_i M_{\langle a_i,b_i,c_i\rangle}\right)$$

In other words, if we can prove a good estimate on the border rank of a direct sum of matrix multiplication tensors, we obtain a bound on ω. Schönhage applied the asymptotic sum inequality to the tensor $M_{\langle 1,4,4\rangle} \oplus M_{\langle 9,1,1\rangle}$ to obtain $\omega < 2.55$.

We now introduce the notion of *degeneracy value* of a tensor. Intuitively, a tensor T with high degeneracy value "contains many large matrix multiplication tensors". More precisely, it means that a high Kronecker power of T degenerates to a direct sum of many disjoint matrix multiplication tensors.

Definition 3.1.4 (See [Lan17, Definition 3.4.7.1]). Let $T \in U \otimes V \otimes W$ be a tensor. For $N \in \mathbb{N}$, we define

$$V_{\omega,N}(T) = \sup\left\{\sum_{i=1}^{q} (a_i b_i c_i)^{\omega/3} \mid T^{\boxtimes N} \trianglerighteq \bigoplus_{i=1}^{q} M_{\langle a_i,b_i,c_i\rangle}\right\},$$

where we take the supremum over all possible ways of degenerating $T^{\boxtimes N}$ to a direct sum of matrix multiplication tensors. The *degeneracy value* $V_\omega(T)$ of T is defined as the supremum $\sup_N V_{\omega,N}(T)^{\frac{1}{N}}$.

Remark 3.1.5. Using Fekete's lemma, one can show that the sequence $V_{\omega,N}(T)^{\frac{1}{N}}$ tends to a limit, and that $V_\omega(T) = \lim_{N \to \infty} V_{\omega,N}(T)^{\frac{1}{N}}$. See [AFLG15] for details.

Example 3.1.6. If $T = M_{\langle a,b,c\rangle}$, then $T^{\boxtimes N} = M_{\langle a^N,b^N,c^N\rangle}$, therefore

$$V_\omega(M_{\langle a,b,c\rangle}) = (abc)^{\omega/3}$$

It follows immediately from the definition that $T \trianglerighteq T'$ implies $V_\omega(T) \geq V_\omega(T')$, and that value is supermultiplicative: $V_\omega(T \boxtimes T') \geq V_\omega(T)V_\omega(T')$. It is also not hard to show $V_\omega(T \oplus T') \geq V_\omega(T) + V_\omega(T')$ (superadditivity); see [Lan17, Section 3.4.7] for details. We can rephrase the asymptotic sum inequality in terms of degeneracy value, as follows.

Theorem 3.1.7. $V_\omega(T) \leq \underline{\mathrm{rk}}(T)$.

Proof. If $T^{\boxtimes N} \trianglerighteq \bigoplus_{i=1}^{q} M_{\langle a_i,b_i,c_i\rangle}$, then

$$\sum_{i=1}^{q} (a_i b_i c_i)^{\omega/3} \leq \underline{\mathrm{rk}}\left(\bigoplus_{i=1}^{q} M_{\langle a_i,b_i,c_i\rangle}\right) \leq \underline{\mathrm{rk}}(T^{\boxtimes N}) \leq \underline{\mathrm{rk}}(T)^N,$$

where the first inequality is Theorem 3.1.3. $\qquad\square$

If we can find a tensor T with low border rank and high value, Theorem 3.1.7 yields a bound on ω. The main difficulty is to obtain good lower bounds on the value of a tensor.

Remark 3.1.8. The infimum $\inf_N(\mathrm{rk}(T^{\boxtimes N}))$ is known as the *asymptotic rank* of T. By Fekete's lemma we can replace this infimum by a limit, and in [Bin80] it was proven that the rank can be replaced by border rank. We can strengthen Theorem 3.1.7 by replacing rank with asymptotic rank. In other words, instead of estimating the border rank of T, we can estimate the border rank of a tensor power of T.

3.1.2 Combinatorial degeneration

Certain degenerations of tensors can be described in a combinatorial way. Let $T \in U \otimes V \otimes W$ be a tensor. A *blocking* D of T is given by decompositions $U = \bigoplus_{i \in I} U_i$, $V = \bigoplus_{j \in J} V_j$, $W = \bigoplus_{k \in K} W_k$. These induce a unique decomposition

$$T = \sum_{(i,j,k) \in I \times J \times K} T_{(i,j,k)}.$$

The *support* of D, denoted $\mathrm{supp}_D T$ consists of all triples $(i, j, k) \in I \times J \times K$ for which $T_{(i,j,k)} \neq 0$.

Remark 3.1.9. If D' is a blocking of another tensor $T' \in U' \otimes V' \otimes W'$, given by $U' = \bigoplus_{i' \in I'} U'_{i'}$, $V' = \bigoplus_{j' \in J'} V'_{j'}$, $W' = \bigoplus_{k' \in K'} W'_{k'}$, then the Kronecker product $T \boxtimes T'$ has a blocking $D \boxtimes D'$ given by

$$U \otimes U' = \bigoplus_{(i,i') \in I \times I'} U_i \otimes U'_{i'},$$

and similar for $V \otimes V'$ and $W \otimes W'$. It holds that

$$\mathrm{supp}_{D \boxtimes D'} T \boxtimes T' = \mathrm{supp}_D T \times \mathrm{supp}_{D'} T' \subseteq (I \times I') \times (J \times J') \times (K \times K').$$

Definition 3.1.10 ([BCS97, (15.29)]). Let I, J, K be finite sets. If $\Psi \subseteq \Phi \subseteq I \times J \times K$, we call Ψ a *combinatorial degeneration* of Φ, written $\Psi \trianglelefteq \Phi$, if there exist functions $\alpha : I \to \mathbb{Z}$, $\beta : J \to \mathbb{Z}$, $\gamma : K \to \mathbb{Z}$, such that $\alpha(i) + \beta(j) + \gamma(k) = 0$ whenever $(i, j, k) \in \Psi$, and $\alpha(i) + \beta(j) + \gamma(k) > 0$ whenever $(i, j, k) \in \Phi \setminus \Psi$.

Proposition 3.1.11 ([BCS97, (15.30)]). *Let D be a blocking of T, with components $T_{(i,j,k)}$, $(i,j,k) \in \mathrm{supp}_D T \subseteq I \times J \times K$. Let Ψ be a combinatorial degeneration of $\mathrm{supp}_D T$. Then*

$$\sum_{(i,j,k) \in \Psi} T_{(i,j,k)} \trianglelefteq T,$$

where $\trianglelefteq$ denotes the usual tensor degeneration.

Next, we need to introduce diagonals and tight sets.

Definition 3.1.12. Let I, J, K be finite sets and $\Delta \subseteq I \times J \times K$.

- We say Δ is a *diagonal*, if the three projections $\Delta \to I$, $\Delta \to J$, $\Delta \to K$ are injective.

- We say Δ is *tight*, if there are injections $\alpha : I \to \mathbb{Z}^r$, $\beta : J \to \mathbb{Z}^r$, $\gamma : K \to \mathbb{Z}^r$ s.t. $\alpha(i) + \beta(j) + \gamma(k) = 0$ for all $(i, j, k) \in \Delta$. If moreover α, β, γ can be chosen such that their images are contained in $\{-b, -b+1, \ldots, b-1, b\}^r$, we say Δ is *b-tight*.

Example 3.1.13. Let $I = \{0, 1, \ldots, n_1\}$, $J = \{0, 1, \ldots, n_2\}$, $K = \{0, 1, \ldots, n_3\}$, and let $\Delta \subseteq I \times J \times K$. Suppose there is a constant c such that $i + j + k = c$ for every $(i, j, k) \in \Delta$. Then Δ is tight: take for example $r = 1$ and define $\alpha(i) = i$, $\beta(j) = j$, $\gamma(k) = c - k$.

Remark 3.1.14.   • Trivially, subsets of b-tight sets are b-tight.

• Let $\Phi_1 \subseteq I_1 \times J_1 \times K_1$ and $\Phi_2 \subseteq I_2 \times J_2 \times K_2$ be b-tight. Then $\Phi_1 \times \Phi_2 \subseteq (I_1 \times I_2) \times (J_1 \times J_2) \times (K_1 \times K_2)$ is b-tight. To see this, just consider the obvious injections $I_1 \times I_2 \to \mathbb{Z}^{r_1 + r_2}$, $J_1 \times J_2 \to \mathbb{Z}^{r_1 + r_2}$, $K_1 \times K_2 \to \mathbb{Z}^{r_1 + r_2}$.

Note that if $\operatorname{supp}_D T$ is a diagonal, then

$$T = \bigoplus_{(i,j,k) \in \operatorname{supp}_D T} T_{(i,j,k)}.$$

The idea behind tight sets is that they are not quite diagonals, but they contain large diagonals that are combinatorial degenerations. This is made precise in the theorem below, which goes back to Coppersmith and Winograd [CW90]. The proof uses probabilistic methods. We call a subset $\Phi \subseteq I \times J \times K$ *balanced*, if the projection $p_I : \Phi \to I$ is surjective, with all fibers of equal cardinality $|\Phi|/I$, and similar for the other projections p_J, p_K. The balancedness assumption in the following theorem is not really necessary, but it simplifies the statement.

Theorem 3.1.15 ([BCS97, (15.39), attributed to Strassen]). *There exists a constant C_b, only depending on b, such that every b-tight balanced subset $\Phi \subseteq I \times J \times K$ contains a diagonal of size at least $C_b \cdot \min\{|I|, |J|, |K|\}$.*

3.1.3   The laser method

We present a variation of [BCS97, Theorem 15.41]. The idea is that if we have a tensor T with a tight blocking whose blocks have high value, then T also has high value. Our Theorem 3.1.20 is a strengthening of [BCS97, Theorem 15.41] in two ways: first, we don't assume that the blocks are matrix multiplication tensors. Second, instead of assuming a lower bound on the value of the blocks, we tensor each block with some of its permutations and assume a bound on the value of those products. More precisely: the symmetric group $\mathfrak{S}_3$ acts on $U \otimes V \otimes W$ by permuting the factors. Note that for $\sigma \in \mathfrak{S}_3$, the tensor $\sigma T \in \sigma U \otimes \sigma V \otimes \sigma W$ has a blocking σD, with components $(\sigma T)_{(\sigma(i), \sigma(j), \sigma(k))} = \sigma(T_{(i,j,k)})$. We fix a subgroup $G \subseteq \mathfrak{S}_3$. Usually, G will be the cyclic group $\mathbb{Z}/3\mathbb{Z} \subseteq \mathfrak{S}_3$. For any tensor T, we will write denote its *symmetrization* $\boxtimes_{\sigma \in G} \sigma T$ by $\tilde{T}$. Note that $V_\omega(\tilde{T}) \geq (V_\omega(T))^{|G|}$, by supermultiplicativity. However, we need to pay a price for the above strengthenings: we need to assume that $\operatorname{supp}_D T$ is *reconstructible*, see Definition 3.1.16 below.

We will consider probability distributions on a finite set $\Phi \subseteq I \times J \times K$. Such a probablility distribution is simply a map $P : \Phi \to [0,1] \subset \mathbb{R}$ such that $\sum_{x \in \Phi} P(x) = 1$. The marginal distribution $P_I : I \to [0,1]$ is defined by

$$P_I(i) = \sum_{j,k \mid (i,j,k) \in \Phi} P(i,j,k)$$

and similar for P_J and P_K.

Definition 3.1.16. We say that a subset $\Phi \subseteq I \times J \times K$ is *reconstructible*, if every probability distribution on Φ is uniquely determined by its 3 marginal distributions.

Example 3.1.17. Let $I = J = K = \{0,1,2\}$, and take

$$\Phi = \{(2,0,0),(0,2,0),(0,0,2),(1,1,0),(1,0,1),(0,1,1)\}.$$

Then Φ is tight (by Example 3.1.13) and reconstructible: if P is a probability distribution on Φ, then the equalities $P(2,0,0) = P_I(2)$, $P(0,1,1) = P_I(0) - P_J(2) - P_K(2),\dots$ allow us to reconstruct P from the marginal distributions P_I, P_J, P_K.

Example 3.1.18. Let $I = J = K = \{0,1,2\}$, and take

$$\Phi = \{(2,1,0),(1,2,0),(2,0,1),(1,0,2),(0,2,1),(0,1,2)\}.$$

Then Φ is again tight by Example 3.1.13, but not reconstructible. For instance, the uniform distribution $P(x,y,z) = \frac{1}{6}$ and the distrubution $P(0,1,2) = P(2,0,1) = P(1,2,0) = \frac{1}{3}$ have the same marginals.

For P a probability distribution on a finite set S, its *entropy* $H(P)$ is defined by $H(P) = -\sum_i P(i) \log P(i)$. The entropy will enter in the proof of Theorem 3.1.20 through the following Lemma, which is an easy consequence of Stirling's formula.

Lemma 3.1.19 (See [BCS97, (15.40)]). *Fix a finite set S. There exists a sequence ρ_N, with $\lim_{N \to \infty} \rho_N = 0$, such that for every rational probability distribution P on S that can be written as $P(i) = Q(i)/N$ for some $Q : S \to \mathbb{N}$, it holds that*

$$\left| \frac{1}{N} \log \binom{N}{Q} - H(P) \right| \leq \rho_N.$$

Here, $\binom{N}{Q}$ stands for the appropriate multinomial coefficient, and $\log$ is the logarithm in base e. Explicitly: if $S = \{1,2,\dots,n\}$, then $\binom{N}{Q} = \binom{N}{Q(1),\dots,Q(n)}$.

We are now ready to state our version of the Laser method.

Theorem 3.1.20. *Let $T \in U \otimes V \otimes W$, and let D be a blocking of T, indexed by $I \times J \times K$. Assume that $\operatorname{supp}_D T$ is tight and reconstructible. Let P be any probability distribution on $\operatorname{supp}_D T$ and let $G \subseteq \mathfrak{S}_3$ be a subgroup. Then*

$$\log V_\omega(\tilde{T}) \geq \min_{L \in \{I,J,K\}} \sum_{\sigma \in G} H(P_{\sigma L}) + \sum_{\operatorname{supp}_D T} P(i,j,k) \log V_\omega(\widetilde{T_{(i,j,k)}}), \qquad (3.1.1)$$

where $\tilde{T}$ denotes the symmetrization with respect to G.

Proof. Assume that D is b-tight. We will assume that P is a rational probablity distribution, i.e. $P(i,j,k) \in \mathbb{Q}$ for all (i,j,k). Since a general probability distribution can be approximated by rational ones, this suffices (see also the proof of [BCS97, (15,41)]). There is a map $Q : \operatorname{supp}_D T \to \mathbb{N}$ and an $N \in \mathbb{N}$ such that $P(i,j,k) = Q(i,j,k)/N$ for all (i,j,k). Let $I_Q \subseteq I^N$ consist of all sequences in which the element i appears exactly $Q_I(i) := N \cdot P_I(i) = \sum_{j,k} Q(i,j,k)$ times for all i. Note that $|I_Q| = \binom{N}{Q_I}$. We define $J_Q \subseteq J^N$ and $K_Q \subseteq K^N$ analogously.

Write $\tilde{T} := \boxtimes_{\sigma \in G}(\sigma T)$, and consider the tensor

$$\tilde{T}^{\boxtimes N} = \boxtimes_{\sigma \in G}(\sigma T)^{\boxtimes N} \in \left(\boxtimes_{\sigma \in G}(\sigma U)^{\boxtimes N} \right) \otimes \left(\boxtimes_{\sigma \in G}(\sigma V)^{\boxtimes N} \right) \otimes \left(\boxtimes_{\sigma \in G}(\sigma W)^{\boxtimes N} \right).$$

By Remark 3.1.9, $\tilde{T}$ has a blocking $\tilde{D} := \boxtimes_{\sigma \in G}(\sigma D)^{\boxtimes N}$, with support

$$\operatorname{supp}_{\tilde{D}} \tilde{T}^{\boxtimes N} = \prod_{\sigma \in G} (\operatorname{supp}_{\sigma D} \sigma T)^N \subseteq \left(\prod_{\sigma \in G} (\sigma I)^N \right) \times \left(\prod_{\sigma \in G} (\sigma J)^N \right) \times \left(\prod_{\sigma \in G} (\sigma K)^N \right),$$

which is again b-tight. We define

$$\Phi := \left((\prod_{\sigma \in G} \sigma I_Q) \times (\prod_{\sigma \in G} \sigma J_Q) \times (\prod_{\sigma \in G} \sigma K_Q) \right) \cap \operatorname{supp}_{\tilde{D}} \tilde{T}^{\boxtimes N}.$$

It trivially holds that $\Phi \trianglelefteq \operatorname{supp}_{\tilde{D}} \tilde{T}^{\boxtimes N}$.

Let $(x,y,z) \in (\prod_{\sigma \in G} (\sigma I)^N) \times (\prod_{\sigma \in G} (\sigma J)^N) \times (\prod_{\sigma \in G} (\sigma K)^N)$, and write

$$(x,y,z) = \left((i_{\sigma,\ell})_{\sigma \in G, 1 \leq \ell \leq N}, (j_{\sigma,\ell})_{\sigma \in G, 1 \leq \ell \leq N}, (k_{\sigma,\ell})_{\sigma \in G, 1 \leq \ell \leq N} \right).$$

From our reconstructibility assumption, it follows that $(x,y,z) \in \Phi$ if and only if for every $\sigma \in G$ and $(\sigma(i), \sigma(j), \sigma(k)) \in \operatorname{supp}_{\sigma D} \sigma T$, there are exactly $Q(i,j,k)$ indices ℓ for which $(i_{\sigma,\ell}, j_{\sigma,\ell}, k_{\sigma,\ell}) = (\sigma(i), \sigma(j), \sigma(k))$. We find

$$\tilde{T}^{\boxtimes N}_{(x,y,z)} = \boxtimes_{\sigma,\ell} (\sigma T)_{(i_{\sigma,\ell}, j_{\sigma,\ell}, k_{\sigma,\ell})} = \boxtimes_{\substack{(i,j,k) \, \in \, \operatorname{supp}_D T \\ \sigma \, \in \, G}} \sigma(T_{(i,j,k)})^{Q(i,j,k)}$$

$$= \boxtimes_{(i,j,k) \in \operatorname{supp}_D T} (\widetilde{T_{(i,j,k)}})^{Q(i,j,k)},$$

hence by supermultiplicativity:

$$V_\omega(\tilde{T}^{\boxtimes N}_{(x,y,z)}) \geq \prod_{(i,j,k)\in\mathrm{supp}_D T} V_\omega(\widetilde{T_{(i,j,k)}})^{Q(i,j,k)}.$$

We now apply Theorem 3.1.15 to the balanced b-tight subset $\Phi \subseteq (\prod_\sigma \sigma I_Q) \times (\prod_\sigma \sigma J_Q) \times (\prod_\sigma \sigma K_Q)$, and we find a diagonal $\Delta \trianglelefteq \Phi$, such that $|\Delta| \geq C_b \cdot \min_{L\in\{I,J,K\}} \prod_\sigma |\sigma L_Q|$. Since $\Phi \trianglelefteq \mathrm{supp}_{\tilde{D}} \tilde{T}^{\boxtimes N}$, we find $\Delta \trianglelefteq \mathrm{supp}_{\tilde{D}} \tilde{T}^{\boxtimes N}$. So by applying 3.1.11, we find

$$\bigoplus_{(x,y,z)\in\Delta} \tilde{T}^{\boxtimes N}_{(x,y,z)} \trianglelefteq \tilde{T}^{\boxtimes N}.$$

So we find

$$V_\omega(\tilde{T}) \geq \Big(\sum_{(x,y,z)\in\Delta} V_\omega(\tilde{T}^{\boxtimes N}_{(x,y,z)}) \Big)^{\frac{1}{N}} \geq \Big(|\Delta| \prod_{(i,j,k)\in\mathrm{supp}_D T} V_\omega(\widetilde{T_{(i,j,k)}})^{Q(i,j,k)} \Big)^{\frac{1}{N}}.$$

By taking logarithms, we obtain

$$\log V_\omega(\tilde{T}) \geq \frac{1}{N} \min_{L\in\{I,J,K\}} \log \Big(C_b \cdot \prod_\sigma \binom{N}{Q_{\sigma L}} \Big) + \sum_{\mathrm{supp}_D T} P(i,j,k) \log \Big(V_\omega(\widetilde{T_{(i,j,k)}}) \Big).$$

Now the theorem follows by taking $N \to \infty$ and applying Lemma 3.1.19. $\qquad\square$

Combining Theorems 3.1.7 and 3.1.20, we can obtain bounds on ω. As examples, we rederive Strassen's bound [Str87], and two bounds obtained by Coppersmith and Winograd [CW90]. In all examples, we take $G = \mathbb{Z}/3\mathbb{Z} \subset \mathfrak{S}_3$. Note that in this case, the term $\min_{L\in\{I,J,K\}} \sum_{\sigma\in G} H(P_{\sigma L})$ in (3.1.1) is equal to $H(P_I) + H(P_J) + H(P_K)$.

Example 3.1.21 (Strassen's tensor). Consider the following tensor:

$$T = T_{STR,n} := \sum_{i=1}^n u_0 \otimes v_i \otimes w_i + \sum_{i=1}^n u_i \otimes v_0 \otimes w_i \in U \otimes V \otimes W,$$

where U, V and W have respective bases $\{u_0, u_1, \ldots, u_n\}$, $\{v_0, v_1, \ldots, v_n\}$, and $\{w_1, \ldots, w_n\}$. We consider the block decomposition D with $I = J = \{0,1\}$, $K = \{1\}$, $U_0 = \langle u_0 \rangle$, $U_1 = \langle u_1, \ldots u_n \rangle$, $V_0 = \langle v_0 \rangle$, $V_1 = \langle v_1, \ldots v_n \rangle$, $W_1 = \langle w_1, \ldots, w_n \rangle$. Then $\mathrm{supp}_D T = \{(0,1,1),(1,0,1)\}$ which is clearly tight and reconstructible. Since $T_{(0,1,1)} = \sum_i u_0 \otimes v_i \otimes w_i = M_{\langle 1,1,n \rangle}$, we find that $\widetilde{T_{(0,1,1)}} = M_{\langle n,n,n \rangle}$. Similarly $\widetilde{T_{(1,0,1)}} = M_{\langle n,n,n \rangle}$. Let P be the uniform distribution on $\{(0,1,1),(1,0,1)\}$, then $H(P_1) = H(P_2) = \log(2)$ and $H(P_3) = 0$. Theorem 3.1.20 now gives

$$\log V_\omega(\tilde{T}) \geq 2\log(2) + \log(n^\omega),$$

so that $V_\omega(\tilde{T}) \geq 4n^\omega$. The border rank of T is equal[1] to $n+1$, as can be seen from writing it as a limit

$$\lim_{t \to 0} \frac{1}{t} \Big(\sum_{j=1}^{n} (u_0 + tu_i) \otimes (v_0 + tv_i) \otimes w_i - u_0 \otimes v_0 \otimes (w_1 + \cdots + w_n) \Big).$$

Hence the border rank of $\tilde{T}$ is at most $(n+1)^3$, hence Theorem 3.1.7 yields

$$4n^\omega \leq (n+1)^3.$$

For $n = 5$, this gives Strassen's bound $\omega < 2.48$.

Example 3.1.22 (The small Coppersmith-Winograd tensor). Consider the following tensor:

$$T_{cw,n} := \sum_{i=1}^{n} (u_0 \otimes v_i \otimes w_i + u_i \otimes v_0 \otimes w_i + u_i \otimes v_i \otimes w_0) \in \mathbb{C}^{n+1} \otimes \mathbb{C}^{n+1} \otimes \mathbb{C}^{n+1}.$$

The border rank of $T_{cw,n}$ is equal to $n+2$; see [Lan17, Proposition 3.4.9.1]. We consider the blocking D with $I = J = K = \{0,1\}$, $U_0 = \langle u_0 \rangle$, $U_1 = \langle u_1, \ldots u_n \rangle$, and similar for V and W. Then $\mathrm{supp}_D T = \{(0,1,1),(1,0,1),(1,1,0)\}$ which is clearly tight and reconstructible. As with Strassen's tensor, we find that $\widetilde{T_{(0,1,1)}} = \widetilde{T_{(1,0,1)}} = \widetilde{T_{(1,1,0)}} = M_{\langle n,n,n \rangle}$. Let P be the uniform distribution on $\{(0,1,1),(1,0,1),(1,1,0)\}$, then $H(P_1) = H(P_2) = H(P_3) = \frac{1}{3}\log(\frac{27}{4})$. Theorem 3.1.20 now gives

$$\log V_\omega(\tilde{T}) \geq \log(\frac{27}{4}) + \log(n^\omega),$$

which combined with the asymptotic sum inequality gives and the estimate $\underline{\mathrm{rk}}(\widetilde{T_{cw,n}}) \leq (n+2)^3$ yields

$$\omega \leq \log_n \Big(\frac{4(n+2)^3}{27} \Big).$$

For $n = 8$, this gives the bound $\omega < 2.40364$ from [CW90, Section 6].

Example 3.1.23 (The big Coppersmith-Winograd tensor). Consider the following tensor:

$$T = T_{CW,n} := \sum_{i=1}^{n} (u_0 \otimes v_i \otimes w_i + u_i \otimes v_0 \otimes w_i + u_i \otimes v_i \otimes w_0)$$

$$+ u_0 \otimes v_0 \otimes w_{n+1} + u_0 \otimes v_{n+1} \otimes w_0 + u_{n+1} \otimes v_0 \otimes w_0 \in \mathbb{C}^{n+2} \otimes \mathbb{C}^{n+2} \otimes \mathbb{C}^{n+2}.$$

---

[1] The border rank cannot be lower than $n+1$, as T is a concise tensor, cfr. Section 1.1.

It be shown that $\underline{\mathrm{rk}}(T_{CW,n}) = n + 2$, for example by providing an explicit border rank decomposition (see [Lan17, Exercise 3.4.9.3]). Alternatively, it will follow from Example 3.3.8. We consider the blocking D with $I = J = K = \{0, 1, 2\}$, $U_0 = \langle u_0 \rangle$, $U_1 = \langle u_1, \ldots u_n \rangle$, $U_2 = \langle u_{n+1} \rangle$, and similar for V and W. Then $\mathrm{supp}_D T = \{(0,1,1), (1,0,1), (1,1,0), (2,0,0), (0,2,0), (0,0,2)\}$, which is tight and reconstructible. As in the previous example, we find that $\widetilde{T_{(0,1,1)}} = \widetilde{T_{(1,0,1)}} = \widetilde{T_{(1,1,0)}} = M_{\langle n,n,n \rangle}$. In addition, $\widetilde{T_{(2,0,0)}} = \widetilde{T_{(0,2,0)}} = \widetilde{T_{(0,0,2)}} = M_{\langle 1,1,1 \rangle}$. This time, it is not clear what is the best probability distribution P. However, for fixed n, we can still write down the bound $V_\omega(\widetilde{T_{CW,n}})$ from Theorem 3.1.20 in terms of P, maximize over all probability distributions P, and obtain a bound on ω. The best such bound is obtained by putting $n = 6$, $P((2,0,0)) = P((2,0,0)) = P((2,0,0)) \approx 0.0160$, and $P((0,1,1)) = P((1,0,1)) = P((1,1,0)) \approx 0.3173$. Then we obtain

$$\omega < 2.38719,$$

which is the bound from [CW90, Section 7]. All subsequent bounds on ω ([CW90, Section 8]; [Sto10, Wil12, LG14]) were obtained by analyzing tensors powers of the Coppersmith-Winograd tensor $T_{CW,n}$.

3.2 Plethysm for fast matrix multiplication

The *symmetrized matrix multiplication tensor* SM_n is given by the following polynomial in n^2 variables x_{ij}, $1 \leq i, j \leq n$:

$$SM_n := \sum_{i,j,k=1}^{n} x_{ij} x_{jk} x_{ki}.$$

Equivalently, if X is the matrix with entries x_{ij}, then $SM_n(X) = \mathrm{tr}(X^3)$. The following theorem says that for estimating the value of ω we can, instead of considering the usual matrix multiplication tensor and its (border) rank, consider the symmetrized matrix multiplication tensor and its Waring (border) rank.

Theorem 3.2.1 ([CHI$^+$18, Theorem 1.1]). *The following equalities hold:*

$$\omega = \inf\{\tau \in \mathbb{R} \mid \mathrm{wrk}(SM_n) = O(n^\tau)\} = \inf\{\tau \in \mathbb{R} \mid \underline{\mathrm{wrk}}(SM_n) = O(n^\tau)\}.$$

Now, consider the SL_n-representation $\mathfrak{gl}_n$, which is simply the space of $n \times n$ matrices, with an action of SL_n given by $A \cdot X = A^{-1}XA$ for $A \in SL_n$ and $X \in \mathfrak{gl}_n$. A basis of $\mathfrak{gl}_n$ is given by $\{E_{i,j}\}_{1 \leq i,j \leq n}$, where $E_{i,j}$ is the matrix with a 1 at position (i, j) and 0's at all other positions. After identifying x_{ij} with $E_{i,j}$, the symmetrized matrix multiplication tensor SM_n naturally lives in $S^3(\mathfrak{gl}_n)$. Moreover, it is an *invariant*: $A \cdot SM_n = SM_n$ for all $A \in SL_n$, as can be seen from the equality $\mathrm{tr}((A^{-1}XA)^3) = \mathrm{tr}(X^3)$. Motivated by this, we will study the representation theory of the space $S^3(\mathfrak{gl}_n)$ in more detail.

3.2.1 Preliminaries on representation theory

In this subsection, we will give a very short overview of the representation theory of GL_n and SL_n, mainly intended to fix notation. We refer the reader also to [Stu08, Chapter 4] for a brief introduction, and to [FH91] for a detailed account.

Let V be an n-dimensional vector space, and let $GL_n = GL(V)$ be the group of linear automorphisms of V. A representation of GL_n is given by a (finite-dimensional) vector space W and a morphism $\varphi : GL_n \to GL(W)$ of linear algebraic groups (i.e. φ is both a morphism of groups and a morphism of varieties). For $A \in GL_n$ and $w \in W$, we will denote $\varphi(A)w$ by $A \cdot w$. A representation is called *irreducible* if it has no nontrivial subrepresentations, i.e. there is no subspace $0 \subsetneq W' \subsetneq W$ such that $A \cdot w \in W'$ for all $A \in GL_n$ and $w \in W'$. A fundamental fact in representation theory is that every representation of GL_n (more generally, of any reductive group) is a direct sum of irreducible representations.

The irreducible representations (sometimes abbreviated as "irreps") of GL_n are in bijection with n-tuples $\lambda = [\lambda_1, \ldots, \lambda_n] \in \mathbb{Z}^n$ with $\lambda_1 \geq \cdots \geq \lambda_n$. The associated GL_n-representation is called a *Weyl module of highest weight* λ, and will be denoted by $\mathbb{S}^\lambda V$. If $\lambda_n \geq 0$, then $\mathbb{S}^\lambda V$ is a *polynomial* representation. In this case, λ is typically presented by a *Young diagram* with at most n rows, where the the i'th row has length λ_i. The Weyl module $\mathbb{S}^\lambda V$ can be obtained by applying the Schur functor $\mathbb{S}^\lambda$ to the natural representation V of $GL(V)$ (i.e. $\varphi : GL(V) \to GL(V)$ is the identity), hence the notation. The Schur functors are defined via Young symmetrizers; we will not recall the precise definition here. Here are two important special cases: $\mathbb{S}^{[a]}W$ is the a-th symmetric power $S^a W$, and $\mathbb{S}^{[1,\ldots,1]}W$ (where 1 appears a times) is the exterior power $\bigwedge^a W$.

We will, however, recall where the name "highest weight module" comes from: Fix a torus $T \simeq (\mathbb{C}^*)^n \subset GL(V)$. Such a torus may be identified with the diagonal nondegenerate matrices, after fixing a basis of V. We recall that a torus T acting on any vector space W induces a *weight decomposition*:

$$W = \bigoplus_{\mathbf{a} \in \mathbb{Z}^n} W_{\mathbf{a}},$$

where $(t_1 \ldots, t_n) \in T$ acts on $v \in W_{\mathbf{a}}$ by scaling as follows:

$$(t_1, \ldots, t_n)v = t_1^{a_1} \cdots t_n^{a_n} v.$$

The $\mathbf{a} \in \mathbb{Z}^n$ for which $W_{\mathbf{a}} \neq 0$ are called the *weights* of the weight decomposition; if $w \in W_{\mathbf{a}}$ we call w a *weight vector* of weight $\mathbf{a}$.

Any irreducible $GL(V)$-representation $W = \mathbb{S}^\lambda V$ decomposes as above under the action of T with a one-dimensional component $W_{\lambda_1,\ldots,\lambda_n}$; moreover all other components have a lexicographically smaller weight. This explains the name "Weyl module of highest weight λ." Note that $\mathbb{S}^\lambda V$ is the *unique* irreducible representation with highest weight λ, so we have at least given an implicit definition of $\mathbb{S}^\lambda V$. An alternative description of the polynomial irreducible representations of $GL(V)$ will be mentioned in Part II of this book (the Borel-Weil theorem, see Theorem 5.4.2).

We now turn our attention to the special linear group $SL(V) \subseteq GL(V)$. Every irrep of $GL(V)$ restricts to an irrep of $SL(V)$, and every irrep of $SL(V)$ arises in this way. Moreover, two $GL(V)$-irreps $\mathbb{S}^\lambda V$, $\mathbb{S}^\mu V$ restrict to the same $SL(V)$-irrep if and only if $\lambda - \mu = [c, c, \ldots, c]$ for some c in $\mathbb{Z}$. Hence (by choosing $\lambda_n = 0$), $SL(V)$-irreps are in bijection with $(n-1)$-tuples $\lambda = [\lambda_1, \ldots, \lambda_{n-1}]$ with $\lambda_1 \geq \ldots \geq \lambda_{n-1} \geq 0$, or equivalently with Young diagrams with at most $n-1$ rows. We will also denote them by $\mathbb{S}^\lambda V$.

An arbitrary $SL(V)$- or $GL(V)$-representation W can be written as a direct sum of irreducible representations. More precisely, there are unique multiplicities $m_\lambda \in \mathbb{N}$ such that

$$W = \bigoplus_\lambda W_\lambda \tag{3.2.1}$$

with

$$W_\lambda \cong (\mathbb{S}^\lambda(V))^{\oplus m_\lambda}. \tag{3.2.2}$$

The terms W_λ are called *isotypic components*. The decomposition (3.2.1) is unique, but the isomorphisms (3.2.2) are not unique when $m_\lambda \geq 2$. A vector $w \in W$ is a *heighest weight vector* if $w \in W_\lambda$ for some λ, and w is a weight vector of weight λ. Equivalently: $w \in W_\lambda$ is a weight vector if the isomorphism (3.2.2) can be chosen such that w is the highest weight vector of one of the summands $\mathbb{S}^\lambda V$. The highest weight vectors of weight λ form a vector space of dimension m_λ. To avoid possible confusion, we stress that if $w \in W$ is a heighest weight vector, this does *not* imply that the weight of w is lexicographically maximal among all weight vectors in W; it only implies that w has lexicographically maximal weight *in its isotypic component*.

We can test whether a given weight vector $w \in W$ is a heighest weight vector using *raising operators*. Recall that the Lie algebra $\mathfrak{sl}_n$ of SL_n is the space of traceless $n \times n$ matrices, equipped with the Lie bracket $[X, Y] := XY - YX$. An action of SL_n on W determines an action of $\mathfrak{sl}_n$ on W by $X \cdot_{\mathfrak{sl}_n} w := \lim_{\varepsilon \to 0} \frac{1}{\varepsilon}\big((I + \varepsilon X) \cdot_{SL_n} w - w\big)$. More abstractly: an SL_n-representation is a map $SL_n \to GL(W)$ of linear algebraic groups, and the induced $\mathfrak{sl}_n$-representation is the differential of this map at the identity matrix. The elements $E_{i,i+1} \in \mathfrak{sl}_n$ are called *raising operators*, because they increase the weight of a weight vector. Explicitely: if $w \in W$ is a weight vector of weight $\mathbf{a}$, then $E_{i,i+1} \cdot w$ is a weight vector of weight $\mathbf{a} + e_i - e_{i+1}$. Raising operators will play a role later because of the following result, which is usually taken as the definition of a highest weight vector.

Proposition 3.2.2 (See [FH91, Proposition 14.13]). *A weight vector $w \in W$ is a heighest weight vector if and only if $E_{i,i+1} \cdot w = 0$ for all $i = 1, \ldots, n-1$.*

3.2.2 The plethysm

In this section we describe a general procedure to decompose $S^k(\mathfrak{gl}_n) = S^k(\mathfrak{gl}(V))$ and $S^k(\mathfrak{sl}_n) = S^k(\mathfrak{sl}(V))$ into irreducibles. The proof uses several facts from representation theory, which we first recall.

- The dual $\mathbb{S}^\lambda(V)^* = \mathbb{S}^\lambda(V^*)$ of an irreducible representation is isomorphic to $\mathbb{S}^{\overline{\lambda}}(V)$, where $\overline{\lambda}$ denotes the partition $[\lambda_1, \lambda_1 - \lambda_{n-1}, \ldots, \lambda_1 - \lambda_2]$. See for example [FH91, Exercise 15.50].

- If W_1 and W_2 are SL_n-representations, then Cauchy's formula says that the symmetric powers of their tensor product decompose as follows [FH91, Exercise 6.11]:

$$S^k(W_1 \otimes W_2) \cong \bigoplus_{\lambda \vdash k} \mathbb{S}^\lambda(W_1) \otimes \mathbb{S}^\lambda(W_2),$$

  where the sum is over all partitions of k of length at most n.

- The decomposition of a tensor product of irreducible representations is given by the Littlewood-Richardson rule

$$\mathbb{S}^\lambda V \otimes \mathbb{S}^\mu V \cong \bigoplus_\nu N^\nu_{\lambda\mu} \mathbb{S}^\nu V.$$

  The *Littlewood-Richardson coefficients* $N^\nu_{\lambda\mu}$, admit a combinatorial description, which can be found in [FH91, Appendix A.1].

Theorem 3.2.3. *It holds that*

$$S^k(\mathfrak{gl}(V)) \cong \bigoplus_{\lambda \vdash k} \bigoplus_\nu N^\nu_{\lambda\overline{\lambda}} \mathbb{S}^\nu V \tag{3.2.3}$$

as $SL(V)$-representations. Here the second summation is over all partitions ν of length at most $n-1$, $N^\nu_{\lambda\mu}$ are the Littlewood-Richardson coefficients, and $\overline{\lambda} = [\lambda_1, \lambda_1 - \lambda_{n-1}, \ldots, \lambda_1 - \lambda_2]$.

Proof. Note that $\mathfrak{gl}(V) \cong V \otimes V^*$ as $SL(V)$-representations. So

$$S^k(\mathfrak{gl}(V)) \cong S^k(V \otimes V^*) \cong \bigoplus_{\lambda \vdash k} \mathbb{S}^\lambda V \otimes \mathbb{S}^\lambda(V)^*$$

$$\cong \bigoplus_{\lambda \vdash k} \mathbb{S}^\lambda V \otimes \mathbb{S}^{\overline{\lambda}} V \cong \bigoplus_{\lambda \vdash k} \bigoplus_\nu N^\nu_{\lambda\overline{\lambda}} \mathbb{S}^\nu V.$$

The second isomorphism holds by Cauchy's formula; the fourth one is precisely the Littlewood-Richardson rule. $\square$

To compute the decomposition of $S^k(\mathfrak{sl}_n)$, we simply note that

$$S^k(\mathfrak{gl}_n) \cong S^k(\mathfrak{sl}_n \oplus \mathbb{C}) \cong \mathbb{C} \oplus \bigoplus_{i=1}^{k} S^i(\mathfrak{sl}_n).$$

This allows us to compute the decomposition of $S^k(\mathfrak{sl}_n)$ inductively.

As a corollary we present an explicit decomposition in the case $k = 3$. Computing the Littlewood-Richardson coefficients in (3.2.3) gives us the decomposition of $S^3(\mathfrak{gl}_n)$ (resp. $S^3(\mathfrak{sl}_n)$) into irreducibles. We present these in Table 3.1: the first column lists the highest weights λ of the occurring irreducible representations $\mathbb{S}^\lambda V$. To be more precise: the first column actually shows the highest weights when we view $S^3(\mathfrak{gl}_n)$ (resp. $S^3(\mathfrak{sl}_n)$) as a GL_n-representation. (Recall that weights of GL_n are n-tuples $[\lambda_1, \ldots, \lambda_n] \in \mathbb{Z}^n$ with $\lambda_1 \geq \ldots \geq \lambda_n$. The corresponding SL_n-weight is then $[\lambda_1 - \lambda_n, \ldots, \lambda_{n-1} - \lambda_n]$.) The second and third column list the multiplicities of the irreducibles in $S^3(\mathfrak{gl}_n)$ resp. $S^3(\mathfrak{sl}_n)$. We also list the dimensions of the occurring irreducible representations $\mathbb{S}^\lambda V$, as well as the dimensions of the projective homogeneous varieties contained in $\mathbb{P}(\mathbb{S}^\lambda V)$ (see below).

Table 3.1: Irreducible components of $S^3(\mathfrak{gl}_n)$ and $S^3(\mathfrak{sl}_n)$

Highest weight	$S^3(\mathfrak{gl}_n)$	$S^3(\mathfrak{sl}_n)$	Dimension	Variety
$[0, \ldots, 0]$	3	1	1	0
$[1, 0, \ldots, 0, -1]$	4	2	$n^2 - 1$	$2n - 3$
$[2, 0, \ldots, 0, -2]$	2	1	$\frac{(n-1)n^2(n+3)}{4}$	$2n - 3$
$[3, 0, \ldots, 0, -3]$	1	1	$\frac{(n-1)n^2(n+1)^2(n+5)}{36}$	$2n - 3$
$[1, 1, 0, \ldots, 0, -1, -1]$	2	1	$\frac{(n-3)n^2(n+1)}{4}$	$4n - 12$
$[2, 0, \ldots, 0, -1, -1]$	1	1	$\frac{(n-2)(n-1)(n+1)(n+2)}{4}$	$3n - 7$
$[1, 1, 0, \ldots, 0, -2]$	1	1	$\frac{(n-2)(n-1)(n+1)(n+2)}{4}$	$3n - 7$
$[2, 1, 0, \ldots, 0, -1, -2]$	1	1	$\frac{(n-3)(n-1)^2(n+1)^2(n+3)}{9}$	$4n - 10$
$[1, 1, 1, 0, \ldots, 0, -1, -1, -1]$	1	1	$\frac{(n-5)(n-1)^2 n^2(n+1)}{36}$	$6n - 27$

Homogeneous varieties

Let V be an irreducible representation of a semisimple Lie group G. Then $\mathbb{P}V$ has a unique closed G-orbit X, which is the orbit of the highest weight vector in $\mathbb{P}V$ under the action of G. The projective variety X is isomorphic to G/P, where P is a parabolic subgroup. We call these varieties homogeneous varieties or partial flag varieties. See also Section 5.4, in particular Theorem 5.4.2.

In our case $G = SL_n$, we can compute the dimension of X in the following way: Consider the Dynkin diagram of $\mathfrak{sl}_n$, which consists of $n - 1$ dots marked

62

1 to $n-1$, and the Young diagram λ associated to the representation V. For every $j \in \{1, \ldots, n-1\}$, if the Young diagram has at least one column of length j, we remove the dot j from the Dynkin diagram. After removing these dots the Dynkin diagram splits in connected components of size k_i. The dimension of our variety X is then given by

$$\frac{1}{2}\left(n^2 - n - \sum_i \left(k_i^2 + k_i\right)\right).$$

This gives us the last column of Table 3.1.

3.2.3 Highest weight vectors

We now describe highest weight vectors for all irreducible components of $S^3(\mathfrak{gl}_n)$. Note that the vector $E_{i,j}E_{i',j'}E_{i'',j''} \in S^3(\mathfrak{gl}_n)$ has weight $e_i + e_{i'} + e_{i''} - e_j - e_{j'} - e_{j''}$, where e_i is the weight $[0, \ldots, 1, \ldots, 0]$ with a 1 on the i-th position. Furthermore, to check that a weight vector v in some representation V of SL_n is a highest weight vector, it suffices to view V as a representation of the Lie algebra $\mathfrak{sl}_n$ and check that every matrix $E_{i,i+1}$ acts by zero (Proposition 3.2.2). Using this, it is straightforward to check that the vectors listed in Table 3.2 are indeed highest weight vectors.

Table 3.2: Highest weight vectors of $S^3(\mathfrak{gl}_n)$

Weight	Highest Weight Vector
$[0, \ldots, 0]$	III
$[0, \ldots, 0]$	$\sum_{i,j} I E_{i,j} E_{j,i}$
$[0, \ldots, 0]$	$\sum_{i,j,k} E_{i,j} E_{j,k} E_{k,i}$
$[1, 0, \ldots, 0, -1]$	$II E_{1,n}$
$[1, 0, \ldots, 0, -1]$	$\sum_i I E_{1,i} E_{i,n}$
$[1, 0, \ldots, 0, -1]$	$\sum_{i,j} E_{1,n} E_{i,j} E_{j,i}$
$[1, 0, \ldots, 0, -1]$	$\sum_{i,j} E_{1,i} E_{i,j} E_{j,n}$
$[2, 0, \ldots, 0, -2]$	$I E_{1,n} E_{1,n}$
$[2, 0, \ldots, 0, -2]$	$\sum_i E_{1,n} E_{1,i} E_{i,n}$
$[1, 1, 0, \ldots, 0, -2]$	$\sum_i E_{1,n} E_{2,i} E_{i,n} - E_{2,n} E_{1,i} E_{i,n}$
$[2, 0, \ldots, 0, -1, -1]$	$\sum_i E_{1,n} E_{1,i} E_{i,n-1} - E_{1,n-1} E_{1,i} E_{i,n}$
$[1, 1, 0, \ldots, 0, -1, -1]$	$I E_{1,n} E_{2,n-1} - I E_{1,n-1} E_{2,n}$
$[1, 1, 0, \ldots, 0, -1, -1]$	$\sum_i E_{1,n} E_{2,i} E_{i,n-1} - E_{2,n} E_{1,i} E_{i,n-1}$ $-E_{1,n-1} E_{2,i} E_{i,n} + E_{2,n-1} E_{1,i} E_{i,n}$
$[3, 0, \ldots, 0, -3]$	$E_{1,n} E_{1,n} E_{1,n}$
$[2, 1, 0, \ldots, 0, -1, -2]$	$E_{1,n} E_{1,n-1} E_{2,n} - E_{1,n} E_{1,n} E_{2,n-1}$
$[1, 1, 1, 0, \ldots, 0, -1, -1, -1]$	$\sum_{\sigma \in S_3} \operatorname{sgn} \sigma E_{\sigma(1),n} E_{\sigma(2),n-1} E_{\sigma(3),n-2}$

Waring rank and border Waring rank

By Theorem 3.2.1, estimating the (border) Waring rank of the highest weight vector $\sum_{i,j,k} E_{i,j} E_{j,k} E_{k,i}$ is equivalent to determining the exponent ω of matrix multiplication. We will analyze the (border) Waring ranks of other highest weight vectors. We start with the following surprising observation.

Observation 3.2.4. Every highest weight vector with weight different from $[0, \ldots, 0]$ has Waring rank $O(n^2)$. Furthermore the weight space of $[0, \ldots, 0]$ is 3-dimensional: it has a basis consisting of two vectors of Waring rank $O(n^2)$, and the vector $\sum_{i,j,k} E_{i,j} E_{j,k} E_{k,i}$.

Proof. Every highest weight vector in Table 3.2, except for $\sum_{i,j,k} E_{i,j} E_{j,k} E_{k,i}$, is a sum of at most n^2 monomials, and every degree 3 monomial has Waring rank at most 4. $\qquad\square$

Recall the small and big Coppersmith-Winograd tensors $T_{cw,m}$ and $T_{CW,m}$ from Examples 3.1.22 and 3.1.23. They are symmetric tensors in $\mathbb{C}[x_0, \ldots, x_{m+1}]_3$, given by

$$T_{cw,m} = \sum_{i=1}^{m} x_0 x_i^2 \quad \text{and} \quad T_{CW,m} = x_0^2 x_{m+1} + \sum_{i=1}^{m} x_0 x_i^2$$

with border Waring rank equal to $m+2$. An interesting observation is that many of the highest weight vectors listed are, up to a change of variables, equal to $T_{cw,m}$ or $T_{CW,m}$ for some value of m.

Proposition 3.2.5. *The following equalities hold, up to a change of variables:*

$$
\begin{aligned}
IE_{1,n} E_{2,n-1} - IE_{1,n-1} E_{2,n} &= T_{cw,4} \\
E_{1,n} E_{1,n-1} E_{2,n} - E_{1,n} E_{1,n} E_{2,n-1} &= T_{CW,2} \\
\sum_i IE_{1,i} E_{i,n} &= T_{cw,2n-2} \\
\sum_i E_{1,n} E_{1,i} E_{i,n} &= T_{CW,2n-4} \\
\sum_{i,j} E_{1,n} E_{i,j} E_{j,i} &= T_{CW,n^2-2}.
\end{aligned}
$$

Proof. We will only prove the fourth equality; the other ones are similar and left to the reader. Note that

$$\sum_{j=1}^{n} E_{1,n} E_{1,j} E_{j,n} = E_{1,n}^2 (E_{1,1} + E_{n,n}) + \sum_{j=2}^{n-2} E_{1,n} E_{1,j} E_{j,n}.$$

If we substitute $E_{1,n} = x_0$, $E_{1,1} + E_{n,n} = x_{2n-3}$, $E_{1,j} = x_{2j-3} + ix_{2j-2}$ and $E_{j,n} = x_{2j-3} - ix_{2j-2}$ for $2 \leq j \leq n-1$ (so that $E_{1,j} E_{j,n} = x_{2j-3}^2 + x_{2j-2}^2$), we obtain the Coppersmith-Winograd tensor $T_{CW,2n-4}$. $\qquad\square$

64

3.3 CW-like tensors via algebraic methods

We now present two approaches for constructing new tensors that are similar to the Coppersmith-Winograd tensor and well-suited for the laser method. More precisely, we are looking for tensors which have low border rank and a (tight and reconstructible) blocking with blocks of high value. The first approach builds upon the work of Landsberg–Michałek [LM17] and Bläser–Lysikov [BL16], the second one is new.

3.3.1 New tensors via smoothable algebras

The main result of this subsection is Theorem 3.3.4, which states that under certain genericity assumptions, a tensor is of minimal border rank if and only if it is the multiplication tensor of a smoothable algebra. We present a self-contained proof of one direction of this Theorem (Proposition 3.3.3). Next, we present an example of a smoothable algebra which gives rise to a tensor that is well-suited for the laser method.

Let $T \in U \otimes V \otimes W$ be a tensor, and assume that T is concise; in particular the map $\phi_T : W^* \to U \otimes V$ is an injection, and $\underline{\mathrm{rk}}(T) \geq \dim W$. Let $L_T \subseteq U \otimes V$ be the image of ϕ_T. We will write $\dim W = \dim L_T = n$. We recall two standard results, which state that we can get information about the rank and border rank of T by studying the space L_T.

Proposition 3.3.1 ([Lan12, Theorem 3.1.1.1]). *The rank of T is equal to n if and only if L_T is spanned by rank one matrices.*

Proof. If $T = \sum_{i=1}^n u_i \otimes v_i \otimes w_i$ then the w_i form a basis of W, and the image under ϕ_T of the dual basis consists of the rank one matrices $u_i \otimes v_i$, which span L_T. Conversely, if L_T is spanned by n rank one matrices $u_i \otimes v_i$, then their preimages under ϕ_T form a basis of W^*. Let $\{w_1, \ldots, w_n\} \subset W$ be the dual basis, then $T = \sum_{i=1}^n u_i \otimes v_i \otimes w_i$. $\qquad\square$

Proposition 3.3.2 (See [LM17, Proposition 2.1]). *If $\mathbb{P}(L_T)$ is the linear span of a smoothable 0-dimensional length n subscheme of the variety of rank one matrices in $\mathbb{P}(U \otimes V)$, then T is of minimal border rank.*

Proof. By assumption, we have a flat family of smooth length n subschemes $R_t \subseteq \mathbb{P}(U \otimes V)$ such that $\mathbb{P}(L_T) = \langle \lim_{t \to 0} R_t \rangle \subseteq \lim_{t \to 0} \langle R_t \rangle$, where the last inclusion is Proposition 1.2.4. This last inclusion is actually an equality, for dimension reasons. In other words, there are families $u_i(t) \otimes v_i(t)$ of rank one matrices such that $L_T = \lim_{t \to 0} L_t$, where $L_t := \langle u_1(t) \otimes v_1(t), \ldots, u_n(t) \otimes v_n(t) \rangle$. By choosing a basis $b_1, \ldots, b_n$ of L_T, and lifting every b_i to a family $b_i(t)$ with $b_i(t) \in L_t$, we can lift the map $\phi_T : W^* \hookrightarrow U \otimes V$ with image L_T to a family of maps $\phi_t : W^* \hookrightarrow U \otimes V$ with images L_t. Write T_t for the tensor corresponding to L_t. Then by Proposition 3.3.1, T_t is a rank n tensor, and $\lim_{t \to 0} T_t = T$. $\qquad\square$

We are now ready to prove the main result of this section, originally due to Bläser and Lysikov [BL16], which says that we can use finite-dimensional algebras to construct tensors of minimal border rank. By convention, all algebras are unital, associative and commutative finite dimensional $\mathbb{C}$-algebras. Such an algebra is called *smoothable* if $\mathrm{Spec}(A)$ is a smoothable scheme. Its multiplication map $V \otimes V \to V$ (where V is the underlying vector space of A) can be seen as a tensor $T_A \in V^* \otimes V^* \otimes V$.

Proposition 3.3.3. *If A is a smoothable algebra, then the multiplication map of A is a tensor of minimal border rank.*

Proof. Let $A = (V, *)$ be an n-dimensional (unital, associative, and commutative) $\mathbb{C}$-algebra. By choosing a vector space basis $\{a_0 = 1, a_1, \ldots, a_{n-1}\}$ of A (with $a_0 = 1$ is the unit of A), we can write $A = \mathbb{C}[a_1, \ldots, a_{n-1}]/I$, where $I = \langle a_i a_j - a_i * a_j \mid 1 \leq i \leq j < n \rangle$. The space L_{T_A} is the image of the dual multiplication map

$$V^* \to V^* \otimes V^* : \sum_i \lambda_i \alpha_i \mapsto \sum_{j,k} \left(\sum_i c^i_{j,k} \lambda_i \right) \alpha_j \otimes \alpha_k,$$

where $\{\alpha_0, \ldots, \alpha_{n-1}\}$ is the dual basis to $\{a_0 = 1, a_1, \ldots, a_{n-1}\}$, and $c^i_{j,k}$ are the structure constants, defined by $a_j * a_k = \sum_i c^i_{j,k} a_i$. Let $X_1 \subseteq \mathbb{P}(V^* \otimes V^*)$ be the variety of rank one matrices, whose defining equations are given by 2×2 minors. Let Y be the scheme-theoretic intersection $\mathbb{P}(L_{T_A}) \cap X_1$. Viewing Y as a subscheme of $\mathbb{P}(V^*)$, its defining ideal is generated by

$$\left(\sum_i c^i_{j,k} \lambda_i \right) \left(\sum_i c^i_{j',k'} \lambda_i \right) - \left(\sum_i c^i_{j',k} \lambda_i \right) \left(\sum_i c^i_{j,k'} \lambda_i \right)$$

for all i, j. By using the fact that the $c^i_{j,k}$ are the structure constants of an algebra, we find that the elements $\sum_i c^i_{j,k} \lambda_0 \lambda_i - \lambda_j \lambda_k$ already generate the ideal. Thus, we can identify Y with $\mathrm{Spec}(A)$. The linear span of $\mathrm{Spec}(A) \subseteq \mathbb{P}(V^* \otimes V^*)$ is $(n-1)$-dimensional and contained in $\mathbb{P}(L_{T_A})$, so it is equal to $\mathbb{P}(L_{T_A})$. The result now follows from Proposition 3.3.2. $\qquad \square$

In fact, as shown by Bläser and Lysikov, the converse holds as well, under certain genericity assumptions.

Theorem 3.3.4 ([BL16, Corollary 3.6]). *Let $T \in V_1 \otimes V_2 \otimes V_3$ be a tensor, with $\dim V_1 = \dim V_2 = \dim V_3$ and assume that the maps $V_1^* \to V_2 \otimes V_3$ and $V_2^* \to V_1 \otimes V_3$ contain a full rank matrix in their image. Then T has minimal border rank if and only if the map $T \in V_1^* \otimes V_2^* \to V_3$ is the multiplication map of a smoothable algebra.*

Hence, one way of constructing tensors that are suitable for the laser method is to consider finite-dimensional $\mathbb{C}$-algebras that are known to be smoothable. In general, checking smoothability is a hard task, but there are several sufficient conditions in the literature for an algebra to be smoothable. In particular, we will use the following result.

Definition 3.3.5. If A is a local algebra, with maximal ideal $\mathfrak{m}$, its Hilbert function is defined by $h_A(i) = \dim \mathfrak{m}^i/\mathfrak{m}^{i+1}$. We will denote the Hilbert function of a local algebra by a sequence $(h_A(0), h_A(1), \ldots)$, where we leave out trailing zeroes. If A is local and graded, then $h_A(i) = \dim A_i$, the dimension of the degree i part of A.

Remark 3.3.6. The Hilbert function as defined above is sometimes called the *local Hilbert function*. There is also the general notion of Hilbert function for subschemes of projective space. Although these concepts are related, they are not the same.

Theorem 3.3.7 ([CEVV09, Propositions 4.12 and 4.13]). *If A is a local algebra with Hilbert function $(1, n, 1)$ or $(1, n, 2)$, then A is smoothable.*

Example 3.3.8. The Coppersmith-Winograd tensor arises as the multiplication tensor of a smoothable algebra. Let V be an $(n+2)$-dimensional vector space with basis $\{a_0, \ldots, a_{n+1}\}$, and let $A = (V, *)$ be the algebra with unit a_0 defined by $a_i * a_i = a_{n+1}$ for $i = 1, \ldots, n$, $a_i * a_j = 0$ for $1 \leq i < j \leq n+1$, and $a_{n+1} * a_{n+1} = 0$. Equivalently $A = \mathbb{C}[a_1, \ldots, a_{n+1}]/I$, where I is the ideal generated by the relations above. The multiplication tensor of A is equal to

$$\alpha_0 \otimes \alpha_0 \otimes a_0 + \sum_{i=1}^{n+1}(\alpha_0 \otimes \alpha_i \otimes a_i + \alpha_i \otimes \alpha_0 \otimes a_i) + \sum_{i=1}^{n}(\alpha_i \otimes \alpha_i \otimes a_{n+1}).$$

We recover the tensor $T_{CW,n}$ from Example 3.1.23 by substituting $\alpha_i = u_i$ in the first tensor factor, $\alpha_i = v_i$ in the second, and $a_0 = w_{n+1}$, $a_{n+1} = w_0$, $a_j = w_j$ in the third (where $1 \leq j \leq n$). Now, A is a local graded algebra, with $\mathfrak{m} = (x_1, \ldots, x_{n+1})$ and grading given by $\deg(x_i) = 1$ for $i = 1, \ldots, n$ and $\deg(x_{n+1}) = 2$. So the Hilbert function of A is equal to $(1, n, 1)$. Hence by Theorem 3.3.7 $\mathrm{Spec}(A)$ is a smoothable scheme, and by Proposition 3.3.2 $T_{CW,m}$ is of minimal border rank.

Example 3.3.9. This example is due to Joachim Jelisiejew. Consider the $3m+3$-dimensional algebra $A = \mathbb{C}[a_1, \ldots, a_{3m+2}]/I$, where the ideal I is generated by

- $a_{3m+1} - a_i a_{m+i}$ for $1 \leq i \leq m$,

- $a_{3m+2} - a_i a_{2m+i}$ for $1 \leq i \leq m$,

- and all products $a_i a_j$, $1 \leq i \leq j \leq 3m + 2$ not occuring in one of the above generators.

Its Hilbert function is given by $(1, 3n, 2)$. As in the previous example, we conclude that $\mathrm{Spec}(A)$ is a smoothable scheme, and hence the multiplication tensor T_A is a tensor of minimal border rank $3m + 3$.

Explicitly, T_A is equal to the following tensor in $A^* \otimes A^* \otimes A$ (with $A \cong \mathbb{C}^{3m+3}$):

$$T_A = \alpha_0 \otimes \alpha_0 \otimes a_0 + \sum_{i=1}^{3m} (\alpha_0 \otimes \alpha_i \otimes a_i + \alpha_i \otimes \alpha_0 \otimes a_i)$$

$$+ \sum_{i=1}^{m} (\alpha_i \otimes \alpha_{m+i} \otimes a_{3m+1} + \alpha_i \otimes \alpha_{2m+i} \otimes a_{3m+2} + \alpha_{m+i} \otimes \alpha_i \otimes a_{3m+1} + \alpha_{2m+i} \otimes \alpha_i \otimes a_{3m+2})$$

$$+ \alpha_0 \otimes \alpha_{3m+1} \otimes a_{3m+1} + \alpha_0 \otimes \alpha_{3m+2} \otimes a_{3m+2} + \alpha_{3m+1} \otimes \alpha_0 \otimes a_{3m+1} + \alpha_{3m+2} \otimes \alpha_0 \otimes a_{3m+2}.$$

The space L_{T_A} consists of all matrices of the form

$$\left(\begin{array}{c|ccc|ccc|ccc|cc}
\lambda_0 & \lambda_1 & \cdots & \lambda_m & \lambda_{m+1} & \cdots & \lambda_{2m} & \lambda_{2m+1} & \cdots & \lambda_{3m} & \lambda_{3m+1} & \lambda_{3m+2} \\ \hline
\lambda_1 & & & & \lambda_{3m+1} & & & \lambda_{3m+2} & & & & \\
\vdots & & & & & \ddots & & & \ddots & & & \\
\lambda_m & & & & & & \lambda_{3m+1} & & & \lambda_{3m+2} & & \\ \hline
\lambda_{m+1} & \lambda_{3m+1} & & & & & & & & & & \\
\vdots & & \ddots & & & & & & & & & \\
\lambda_{2m} & & & \lambda_{3m+1} & & & & & & & & \\ \hline
\lambda_{2m+1} & \lambda_{3m+2} & & & & & & & & & & \\
\vdots & & \ddots & & & & & & & & & \\
\lambda_{3m} & & & \lambda_{3m+2} & & & & & & & & \\ \hline
\lambda_{3m+1} & & & & & & & & & & & \\
\lambda_{3m+2} & & & & & & & & & & &
\end{array}\right).$$

We point out that this looks exactly like the "multiplication table" of A. Also, note that if we take the ideal generated by the 2×2 minors of this matrix and substitute $\lambda_0 = 1$ and $\lambda_i = a_i$ for $i > 0$, we recover I.

We now apply the laser method to T_A. We take the blocking D given by

$$A_0^* = \langle \alpha_0 \rangle, \; A_1^* = \langle \alpha_1, \ldots, \alpha_{3m} \rangle, \; A_2^* = \langle \alpha_{3m+1}, \alpha_{3m+2} \rangle$$

for the first 2 tensor factors, and

$$A_0 = \langle a_{3m+1}, a_{3m+2} \rangle, \; A_1 = \langle a_1, \ldots, a_{3m} \rangle, \; A_2 = \langle a_0 \rangle$$

for the third tensor factor. The support $\mathrm{supp}_D T_A$ is equal to

$$\{(0, 0, 2), (0, 2, 0), (2, 0, 0), (1, 1, 0), (1, 0, 1), (0, 1, 1)\},$$

which is tight and reconstructible. The blocks $T_{A,(2,0,0)} = M_{\langle 2,1,1 \rangle}$, $T_{A,(0,2,0)} = M_{\langle 1,2,1 \rangle}$, $T_{A,(0,0,2)} = M_{\langle 1,1,1 \rangle}$, $T_{A,(0,1,1)} = M_{\langle 1,3m,1 \rangle}$ and $T_{A,(1,0,1)} = M_{\langle 1,1,3m \rangle}$ are all matrix multiplication tensors, and hence their values are known. The final block

$$T_{A,(1,1,0)} = \sum_{i=1}^{m} (\alpha_i \otimes \alpha_{m+i} \otimes a_{3m+1} + \alpha_i \otimes \alpha_{2m+i} \otimes a_{3m+2}$$
$$+ \alpha_{m+i} \otimes \alpha_i \otimes a_{3m+1} + \alpha_{2m+i} \otimes \alpha_i \otimes a_{3m+2}) \quad (3.3.1)$$

can be identified with the Kronecker product

$$(\sum_{j=0}^{m-1} e_j \otimes e_j \otimes 1) \boxtimes (e_0 \otimes e_1 \otimes e_1 + e_0 \otimes e_2 \otimes e_2 + e_1 \otimes e_0 \otimes e_1 + e_2 \otimes e_0 \otimes e_2).$$

The first factor is the matrix multiplication tensor $M_{\langle 1,m,1 \rangle}$, whose symmetrization (with respect to $\mathbb{Z}/3\mathbb{Z} \subset \mathfrak{S}_3$) has value $V_\omega(M_{\langle m,m,m \rangle}) = m^\omega$. The second factor is Strassen's tensor $T_{STR,2}$. By Example 3.1.21, we have $V_\omega(\widetilde{T_{STR,2}}) \geq 4 \cdot 2^\omega$. Hence

$$V_\omega(\widetilde{T_{A,(1,1,0)}}) \geq 4(2m)^\omega.$$

We now apply the laser method Theorem 3.1.20 using this value estimate:

$$3\,\underline{\mathrm{rk}}(T_A) \geq \log V_\omega(\widetilde{T_A}) \geq H(P_1) + H(P_2) + H(P_3) + P(1,1,0) \log\left(4(2m)^\omega\right)$$
$$+ (P(2,0,0) + P(0,2,0)) \log(2^3) + (P(1,0,1) + P(0,1,1)) \log\left((3m)^\omega\right).$$

For fixed m we can obtain a bound on ω, by optimizing over all probability distributions on $\mathrm{supp}_D T_A$. The best bound is obtained by putting $m = 4$: we obtain $\omega < 2.431$.

3.3.2 New tensors via highest weight vectors

Let $n \geq 3$ and write $m = n - 3$. We will focus on the following highest weight vector from Table 3.2:

$$T_{HW,n} := \sum_{i=1}^{n} (E_{1,n}E_{2,i}E_{i,n} - E_{2,n}E_{1,i}E_{i,n}) \in S^3(\mathbb{C}^{n^2}).$$

This is not a concise tensor, but we can make it concise by changing the ambient space: first, rewrite $T_{HW,n}$ as

$$E_{1,n}^2 E_{2,1} - E_{2,n}^2 E_{1,2} + E_{1,n}E_{2,n}(E_{2,2} - E_{1,1}) + \sum_{i=3}^{n-1} (E_{1,n}E_{2,i}E_{i,n} - E_{2,n}E_{1,i}E_{i,n}).$$

For every $i \in \{3 \ldots n-1\}$, we put $E_{i,n} = x_{i-3}$, $E_{2,i} = y_{i-3}$, and $E_{1,i} = z_{i-3}$. Moreover, we put $E_{1,1} - E_{2,2} = b_0$, $E_{1,n} = a_1$, $E_{2,n} = -a_2$, $E_{2,1} = b_1$, $E_{1,2} = -b_2$. Then our tensor becomes

$$T_{HW,m} = a_1 a_2 b_0 + a_1^2 b_1 + a_2^2 b_2 + \sum_{j=1}^{m} (a_1 x_j y_j + a_2 x_j z_j) \in S^3(V) \subseteq V \otimes V \otimes V,$$

where V is the $3m + 5$-dimensional vector space with basis

$$\{a_1, a_2, x_1, \ldots, x_m, y_1, \ldots, y_m, z_1, \ldots, z_m, b_0, b_1, b_2\}.$$

The space L_T consists of all matrices of the form

$$\left(
\begin{array}{cc|ccc|ccc|ccc|ccc}
b_1 & b_0 & y_1 & \cdots & y_m & x_1 & \cdots & x_m & 0 & \cdots & 0 & a_2 & a_1 & 0 \\
b_0 & b_2 & z_1 & \cdots & z_m & 0 & \cdots & 0 & x_1 & \cdots & x_m & a_1 & 0 & a_2 \\
\hline
y_1 & z_1 & & & & a_1 & & & a_2 & & & & & \\
\vdots & \vdots & & & & & \ddots & & & \ddots & & & & \\
y_m & z_m & & & & & & a_1 & & & a_2 & & & \\
\hline
x_1 & 0 & a_1 & & & & & & & & & & & \\
\vdots & \vdots & & \ddots & & & & & & & & & & \\
x_m & 0 & & & a_1 & & & & & & & & & \\
\hline
0 & x_1 & a_2 & & & & & & & & & & & \\
\vdots & \vdots & & \ddots & & & & & & & & & & \\
0 & x_m & & & a_2 & & & & & & & & & \\
\hline
a_2 & a_1 & & & & & & & & & & & & \\
a_1 & 0 & & & & & & & & & & & & \\
0 & a_2 & & & & & & & & & & & &
\end{array}
\right)$$

It is easy to see that $T_{HW,m}$ is concise, i.e. the induced map $V^* \to V \otimes V$ is injective. From this, it follows that $T_{HW,m}$ has border rank at least $\dim V = 3m + 5$. The reason why this highest weight vector in particular is of interest, is that it appears to have minimal border rank.

Conjecture 3.3.10. *The border rank of $T_{HW,m}$ is equal to $3m + 5$.*

Ever stronger, we conjecture that the border *Waring* rank of $T_{HW,m}$ is equal to $3m + 5$. For $m \leq 1$ we have the following exact border decompositions.

Proposition 3.3.11. *For $m = 0$ and $m = 1$, the tensor $T_{HW,m}$ has border Waring rank $3m + 5$.*

Proof. We provide explicit Waring rank approximations in both cases. For the case $m = 0$, let

$$T_{0,t} = 3(a_1 + tb_1)^3 + 6(a_2 + tb_2)^3 + (a_1 - 2a_2)^3$$
$$-3(a_1 - a_2 + tb_0)^3 - (a_1 + a_2 - 3tb_0)^3.$$

Then

$$\lim_{t \to 0} \frac{T_{0,t}}{t} = 36a_1 a_2 b_0 + 9a_1^2 b_1 + 18a_2^2 b_2,$$

which is equal to our tensor $T_{HW,1}$ up to rescaling the variables. For the case $m = 1$, let

$$T_{1,t} = (-a_1 - a_2 + t^3 b_0)^3 + \frac{1}{3}(-a_1 + a_2)^3 + (a_1 - t^2 y_1 + t^3 b_1)^3 +$$

$$\frac{1}{3}(a_1 - a_2 + 3tx_1 - 3t^3 b_0)^3 + \frac{1}{4}(2a_2 - 2tx_1 - t^2 z_1 + t^3 b_2)^3 +$$

$$(-a_1 - 2tx_1 + t^2 y_1)^3 + \frac{1}{4}(-2a_2 + t^2 z_1)^3 + (a_1 + a_2 + tx_1)^3$$

Then

$$\lim_{t \to 0} \frac{T_{1,t}}{t^3} = 12a_1 a_2 b_0 + 3a_1^2 b_1 + 3a_2^2 b_2 + 12a_1 x_1 y_1 + 6a_2 x_1 z_1,$$

which is equal to $T_{HW,1}$ up to rescaling the variables. $\qquad\square$

For $m \geq 2$ there is strong evidence for Conjecture 3.3.10. First, we tried to bound the rank of $T_{HW,m}$, for $m \leq 15$ from below using *Koszul flattenings* [LO13, Proposition 4.1.1], and the obtained lower bound was equal to $3m+5$ in all cases. Second, with the help of Austin Connor, we were able to obtain numerical border rank decompositions of $T_{HW,m}$, for $m \leq 5$.

We now apply the laser method to the tensor $T_m = T_{HW,m}$. The blocking D is given by $V = V_0 \oplus V_1 \oplus V_2$, with

$$V_0 = \langle a_1, a_2 \rangle, V_1 = \langle x_1, \ldots x_m, y_1, \ldots, y_m, z_1, \ldots, z_m \rangle, V_2 = \langle b_0, b_1, b_2 \rangle$$

The block

$$T_{m,(1,1,0)} = \sum_{j=1}^{m} (x_j \otimes y_j \otimes a_1 + x_j \otimes z_j \otimes a_2 + y_j \otimes x_j \otimes a_1 + z_j \otimes x_j \otimes a_2),$$

is precisely the same as the block (3.3.1) from the previous example. Hence its value is at least $4(2m)^\omega$. The same holds for $T_{m,(1,0,1)}$ and $T_{m,(0,1,1)}$.

The value of the other blocks is more difficult to estimate. For instance

$$T_{m,(2,0,0)} = (b_0 \otimes a_1 \otimes a_2 + b_0 \otimes a_2 \otimes a_1 + b_1 \otimes a_1 \otimes a_1 + b_2 \otimes a_2 \otimes a_2).$$

has value at least 2, as substituting $b_0 = 0$ in the first factor yields a direct sum of 2 matrix multiplication tensors $M_{\langle 1,1,1 \rangle}$.

With the above value estimates, Theorem 3.1.20 yields

$$3\,\underline{\mathrm{rk}}(T_m) \geq \log V_\omega(\widetilde{T_m}) \geq H(P_1) + H(P_2) + H(P_3)$$
$$+ (P(2,0,0) + P(0,2,0) + P(0,0,2)) \log(2^3)$$
$$+ (P(1,1,0) + P(1,0,1) + P(0,1,1)) \log\left(4(2m)^\omega\right).$$

Assuming Conjecture 3.3.10, we obtain for every m a bound on ω, by optimizing over all probability distributions on $\mathrm{supp}_D T_m$. The best bound is obtained by putting $m = 7$: then we obtain $\omega < 2.451$.

3.3.3 Outlook

We end this chapter by listing a number of open questions and future directions. First, in Section 3.3.2, we left open the problem of actually proving that $T_{HW,m}$ has minimal border rank. This could potentially be proven either by finding an explicit decomposition, or by using geometric methods like Proposition 3.3.2. We also used a very naive estimates for the value of one of the blocks; improving this estimate would improve the obtained bound $\omega < 2.451$ slightly.

The best known bounds on ω were obtained by an explicit study of higher Kronecker powers of the Coppersmith-Winograd tensor and finding better blockings for those. It might be worthwhile to also study the Kronecker powers of the tensors considered in 3.3.

Finally, the method of Section 3.3.1 can be used to construct many more smoothable tensors. In particular, we plan to undertake a study of algebras with Hilbert series $(1, n, 3)$, and analyze which of these algebras are smoothable and suitable for the laser method.

Chapter 4

The Hessian discriminant

In this chapter, we express the Hessian discriminant of a cubic surface in terms of fundamental invariants. This answers Question 15 of the *27 questions on the cubic surface* posed by Bernd Sturmfels. This chapter is based on joint work with Rodica Dinu [DS20].

A cubic surface in $\mathbb{P}^3$ is the vanishing locus of a degree 3 polynomial

$$f(x_0, x_1, x_2, x_3) := \sum_{0 \le i \le j \le k \le 3} c_{ijk} x_i x_j x_k \in \mathbb{C}[x_0, x_1, x_2, x_3]$$

in 4 variables. The study of cubic surfaces is an important research topic in classical algebraic geometry. Recently, Anna Seigal [Sei19] introduced a new invariant of cubic surfaces called the *Hessian discriminant HD*. It is a homogeneous degree 120 polynomial in the 20 variables c_{ijk}, which is defined as a specialization of the Hurwitz form of the variety of rank 2 symmetric 4×4 matrices.

The rank of a cubic surface $V(f)$ is simply the Waring rank of f; for a general cubic surface the rank is equal to 5. It can be shown that (the vanishing locus of) the Hessian discriminant is precisely the Zariski closure of the set of rank 6 cubic surfaces. There is also a connection between the Hessian discriminant and the more classical study of cubic surfaces: it is related to the singular points of the Hessian surface.

By construction, the Hessian discriminant defines a hypersurface in $\mathbb{P}^{19}$ which is invariant under the action of $PGL(3)$. In other words, HD is an invariant of cubic surfaces. The generators of the invariant ring of cubic surfaces are known [Sal60], so it is natural to ask how to express HD in terms of these fundamental invariants. This was Question 15 in the *27 questions on the cubic surface* [RS19]. The main result of this article provides an answer to this question.

Theorem 4.4.1. $HD = I_{40}^3$, *where I_{40} is the degree 40 Salmon invariant.*

In fact, it is not very hard to deduce this result from known facts about cubic surfaces. However, the required results appear to be quite scattered in the literature. In this chapter, we present a proof that only relies on two very classical results: the classification of cubic surfaces by Schläfli [Sch63], and the computation of the invariant ring by Salmon [Sal60]. We also spend some time explaining connections with Hessian surfaces and with apolar schemes.

The organization of this chapter is as follows: In Section 4.1, we review the definition of the Hurwitz form and the Hessian discriminant. We also explain how to use software to verify whether a given cubic lies on the Hessian discriminant, and explain connections with Hessian surfaces and with apolar schemes. In Section 4.2, we use the classical theory of normal forms for cubic surfaces to decide for every cubic surface outside of a certain codimension 2 locus whether or not it lies on the Hessian discriminant. In Section 4.3, we recall the invariant theory of cubic surfaces, and give a computational proof that the vanishing locus of the invariant I_{40} is the Zariski closure of the set of smooth rank 6 cubic surfaces. Finally, Section 4.4 puts together the results of the preceding two sections to prove Theorem 4.4.1.

4.1 The Hessian discriminant

4.1.1 The Hurwitz form

Let X be an irreducible variety in projective space $\mathbb{P}^n$ of codimension $d \geq 1$ and degree $p \geq 2$. Let $\mathbb{G}(d, \mathbb{P}^n)$ denote the Grassmannian of dimension d subspaces of $\mathbb{P}^n$. Following [Stu17], define $\mathcal{H}_X \subset \mathbb{G}(d, \mathbb{P}^n)$ to be the set of all subspaces L for which $L \cap X$ does not consist of p reduced points. If L is the row space of a matrix $B = (b_{i,j})_{0 \leq i \leq d, 0 \leq j \leq n}$, then the entries $b_{i,j}$ are the *Stiefel coordinates* of L, and the maximal minors of B are the *Plücker coordinates*. One can obtain the *sectional genus* of X by intersecting the variety with a general subspace of dimension $d - 1$ and then taking the arithmetic genus of the obtained curve.

Theorem 4.1.1. *[Stu17, Theorem 1.1] $\mathcal{H}_X$ is an irreducible hypersurface in $\mathbb{G}(d, \mathbb{P}^n)$, defined by an irreducible element Hu_X in the homogeneous coordinate ring of $\mathbb{G}(d, \mathbb{P}^n)$. If X is regular in codimension 1, then the degree of Hu_X in Plücker coordinates equals $2p + 2g - 2$, where g is the sectional genus of X.*

The polynomial Hu_X defined above is called the *Hurwitz form* of X. Interesting examples of Hurwitz forms in computational algebraic geometry can be consulted in [Stu17]. To define the Hessian discriminant, we will need to consider the Hurwitz form of the variety X_2 of symmetric 4×4 matrices of rank at most 2. If we write $\mathbb{P}^9$ for the space of all symmetric 4×4 matrices, then $X_2 \subset \mathbb{P}^9$ is an irreducible subvariety defined by the vanishing of the 3×3 minors. It has dimension 6, degree 10, and sectional genus 6. By Theorem 4.1.1, the Hurwitz form Hu_{X_2} is an irreducible hypersurface of degree 30 in the Plücker coordinates of $\mathbb{G}(3, \mathbb{P}^9)$. In [Stu17], there is an algorithm to compute the polynomial Hu_{X_2}, but it does not finish in a reasonable amount of time in this case.

4.1.2 The Hessian discriminant

For the rest of the chapter, fix a 4-dimensional $\mathbb{C}$-vector space V. Let $\mathcal{C} = V(f)$ be a cubic surface in $\mathbb{P}^3 = \mathbb{P}(V)$, defined by a quaternary cubic

$$f = \sum_{0 \leq i \leq j \leq k \leq 3} c_{ijk} x_i x_j x_k \in \mathbb{C}[x_0, x_1, x_2, x_3]_3 = S^3(V^*).$$

The 20 coefficients c_{ijk} determine a point in $\mathbb{P}(S^3(\mathbb{C}^4)) \cong \mathbb{P}^{19}$. We will use the notions of "cubic surfaces", "quaternary cubics (up to scaling)", and "points in $\mathbb{P}^{19}$" interchangeably. If $\mathcal{C}$ is not a cone over a plane cubic, we can associate to f a 3-plane $H(f)$ in the space $\mathbb{P}^9 = \mathbb{P}(S^2(V^*))$ of symmetric 4×4 matrices. The points of $H(f)$ are called *polar quadrics* of f. There are several equivalent ways to define $H(f)$. We leave it to the reader to check that they are indeed equivalent.

- The Hessian of f is the 4×4 matrix of linear forms whose (i,j)-th entry is $\frac{\partial^2 f}{\partial x_i \partial x_j}$. It defines an injective linear map $i_f : \mathbb{P}^3 \to \mathbb{P}^9$, sending a point $p = [x_0 : x_1 : x_2 : x_3]$ to the Hessian matrix evaluated in that point. We define $H(f)$ to be the image of i_f.

- We can also define $H(f)$ as the linear span of the four partial derivatives $\frac{\partial f}{\partial x_0}, \frac{\partial f}{\partial x_1}, \frac{\partial f}{\partial x_2}, \frac{\partial f}{\partial x_3}$, seen as points in $\mathbb{P}(S^2(V^*))$. Note that these 4 points are well-defined and not coplanar, unless after change of coordinates f is a polynomial in 3 variables. This explains our assumption that $\mathcal{C}$ is not a cone over a plane cubic.

- We can view f as a symmetric three-way tensor $T = (T_{ijk})_{i,j,k}$. (I.e. $c_{ijk} = \lambda T_{ijk}$, where λ is the number of distinct permutations of i, j, k. Then $f = \sum_{i,j,k} T_{ijk} x_i x_j x_k$.) For $m \in \{0, 1, 2, 3\}$, the m-th *slice* of T is defined to be the symmetric matrix obtained by fixing the first index to be m. Then $H(f)$ is the linear span of the four slices of T. From this description we see immediately the Stiefel coordinates of $H(f)$: they are the entries of a 4×10 matrix with colums indexed by pairs (j, k) with $j < k$, whose $i, (j, k)$-th entry is T_{ijk}.

We are now ready to introduce the Hessian discriminant. Recall that Hu_{X_2} is the Hurwitz form of the variety of symmetric 4×4 matrices of rank at most 2.

Definition 4.1.2. The *Hessian discriminant* $HD \in \mathbb{C}[c_{000}, \ldots, c_{333}]_{120}$ is the polynomial obtained by evaluating Hu_{X_2} in the Plücker coordinates of $H(f)$, where f is a general cubic surface.

By construction, the Hessian discriminant vanishes at $f \in \mathbb{P}^{19}$ if and only if $H(f)$ does not intersect the variety of rank 2 matrices in 10 reduced points. Clearly, $V(HD)$ is invariant under linear changes of coordinates. It follows that

HD is invariant under the natural action of $SL(4)$ on $\mathbb{C}[c_{000}, \ldots, c_{333}]$. The following observation connects the Waring rank of cubic forms with the Hessian discriminant.

Observation 4.1.3. (See [Sei19, Section 2.4].) If f has Waring rank at least 6 and defines a smooth cubic surface, then f lies on the Hessian discriminant.

It will be easy to verify this, once we have recalled the normal forms of smooth cubic surfaces in Section 4.2. In [SS19, Corollary 4.4], it is proven that the vanishing locus of the Hessian discriminant is the Zariski closure of the set of all rank 6 cubic surfaces; in particular, all rank ≥ 6 cubics (not just the smooth ones) lie on the Hessian discriminant.

4.1.3   The Hessian surface

In this section, we investigate how the Hessian discriminant is related to the singular locus the Hessian surface of a cubic surface. The determinant locus of the Hessian of f defines a quartic surface $\mathrm{Hess}(f)$ in $\mathbb{P}^3$, called the *Hessian surface* of f. It can be identified with the intersection of $H(f)$ and the variety X_3 of singular 4×4 matrices. Since the singular locus of X_3 is equal to X_2, the locus $H(f) \cap X_2$ of rank 2 matrices in $\mathrm{Hess}(f)$ is contained in the singular locus of $\mathrm{Hess}(f)$. For smooth cubic surfaces, this is an equality.

Proposition 4.1.4. *For a smooth cubic surface, the singular locus* $\mathrm{Sing}(\mathrm{Hess}(f))$ *of its Hessian surface is equal to* $H(f) \cap X_2$.

It follows that a smooth cubic surface lies on the Hessian discriminant if and only if its Hessian surface has strictly less than 10 singular points. This result appears to be well-known, but we were not able to find a complete proof in the literature. In Section 4.2.4, we will give a proof of Proposition 4.1.4 relying on the classification of smooth cubic surfaces. For singular cubic surfaces, the situation is somewhat more subtle: first of all, the following result (which also appears to be folklore) shows that the Hessian surface might have additional singular points.

Proposition 4.1.5. *If a cubic surface f is singular at a point p, then p is a singular point of* $\mathrm{Hess}(f)$.

Proof. Without loss of generality, we may assume that $p = [1 : 0 : 0 : 0]$. Then $f = x_0 g + h$, where g and h are homogenous polynomials of respective degrees 2 and 3 containing only the variables x_1, x_2, x_3. In particular it holds that $\frac{\partial^2 f}{\partial x_0^2} = 0$ and $\frac{\partial^2 f}{\partial x_0 \partial x_i} = \frac{\partial g}{\partial x_i}$ for $i = 1, 2, 3$. Hence the determinant of the Hessian matrix is a homogeneous degree 4 polynomial that does not contain any monomials divisible by x_0^3, which implies that $[1 : 0 : 0 : 0]$ is a singular point of $\mathrm{Hess}(f)$. $\qquad\square$

An explicit example of a cubic surface whose Hessian surface has more than 10 (but finitely many) singularities is the Cayley cubic; see Section 4.2.3. In [Rei, Proposition 4.5], it is shown that for a certain class of (possibly singular) cubic surfaces, the singular locus of the Hessian surface is precisely equal to $(H(f) \cap X_2) \cup \mathrm{Sing}(f)$. However, the following example shows that this is not true for all cubic surfaces.

Example 4.1.6. Consider the cubic surface defined by the equation

$$f := \frac{1}{6}x_0^3 + x_1 x_2 x_3 = 0.$$

The Hessian matrix of f is equal to

$$\begin{pmatrix} x_0 & 0 & 0 & 0 \\ 0 & 0 & x_3 & x_2 \\ 0 & x_3 & 0 & x_1 \\ 0 & x_2 & x_1 & 0 \end{pmatrix},$$

hence its Hessian surface is a union of 4 planes defined by $x_0 x_1 x_2 x_3 = 0$, whose singular points are the 6 lines $x_i = x_j = 0$, $0 \le i < j \le 3$. However, $H(f) \cap X_2$ only consists of the point $[1 : 0 : 0 : 0]$ and the three lines $x_0 = x_i = 0$, $i = 1, 2, 3$, and $\mathrm{Sing}(f)$ consists of 3 points $[0 : 1 : 0 : 0]$, $[0 : 0 : 1 : 0]$, $[0 : 0 : 0 : 1]$.

4.1.4 Computational methods

While computing an expression for the Hessian discriminant is a computationally difficult task, it is easy to verify for a given cubic whether or not the Hessian discriminant vanishes at that cubic: one simply needs to compute the ideal defining the intersection of $H(f) \cap X_2$, and check whether or not it is zero-dimensional and radical. Some code in Macaulay2 [GS] for computing this can be found below:

```
R=QQ[x_0..x_3,z_0..z_9]
X={x_0,x_1,x_2,x_3};
A=genericSymmetricMatrix(R,z_0,4)
I2=minors(3,A)
hessRank2 = f ->(
hess = diff(transpose matrix{X},diff(matrix{X},f));
I=eliminate(X,ideal(flatten entries (A-hess)));
return (I+I2);
)
isOnHessianDiscriminant = f ->(
J=hessRank2(f);
return not ((codim J==9) and (J==radical J));
)
```

```
--Examples:
f=x_0*x_1*x_2+x_0*x_1*x_3+x_0*x_2*x_3+x_1*x_2*x_3
isOnHessianDiscriminant(f)
--false
f=x_0^3+x_1^3+x_2^3+x_3^2*(3*x_0+3*x_1+3*x_2+x_3)
isOnHessianDiscriminant(f)
--true
```

Remark 4.1.7. The above algorithm can also be used to simultaneously compute $H(f) \cap X$ for all f in a family of cubic surfaces. For more details, see Remark 4.2.5, as well as the supplementary code available at https://software.mis.mpg.de/.

4.1.5 Apolarity

There is a beautiful connection between the 3-plane $H(f)$ associated to f and apolar schemes of f. Although not logically necessary for the proof of our main theorem, it can provide some insight in the nature of the singularities of the Hessian surface of a smooth cubic. We will identify the symmetric algebra $S(V)$ of V with the polynomial ring $\mathbb{C}[y_0, y_1, y_2, y_3]$. For every $d, m \in \mathbb{N}$, there is a natural pairing

$$\circ : \mathbb{C}[y_0, y_1, y_2, y_3]_d \times \mathbb{C}[x_0, x_1, x_2, x_3]_m \to \mathbb{C}[x_0, x_1, x_2, x_3]_{m-d}$$

defined by $g \circ f = g(\frac{\partial}{\partial x_0}, \frac{\partial}{\partial x_1}, \frac{\partial}{\partial x_2}, \frac{\partial}{\partial x_3}) f(x_0, x_1, x_2, x_3)$. Note that $H(f)$ can be identified with the image of the map $V \to S^2(V^*) : g \mapsto g \circ f$.

Definition 4.1.8. For f in $\mathbb{C}[x_0, x_1, x_2, x_3]$, we define the *annihilator* of f to be the ideal

$$\mathrm{Ann}(f) = \{g | g \circ f = 0\} \subseteq \mathbb{C}[y_0, y_1, y_2, y_3].$$

If $I \subseteq \mathrm{Ann}(f)$ is a saturated ideal, we say that I is an *apolar ideal* to f, and $V(I)$ is an *apolar scheme* to f. In other words, $Y \subseteq \mathbb{P}(V^*)$ is an apolar scheme to f if every polynomial that vanishes on Y also annihilates f.

Observation 4.1.9. Denote the coordinates on $\mathbb{P}^9 = \mathbb{P}(S^2(V^*))$ by z_{ij}. Then defining equations $\sum_{i \leq j} a_{ij} z_{ij} = 0$ of $H(f)$ are in one-to-one correspondence with degree 2 elements $\sum_{i \leq j} a_{ij} y_i y_j$ of $\mathrm{Ann}(f)$: $\sum_{i \leq j} a_{ij} y_i y_j$ is in $\mathrm{Ann}(f)$ if and only if $\sum_{i \leq j} a_{ij} \frac{\partial^2 f}{\partial x_i \partial x_j} = 0$, if and only if $\sum_{i \leq j} a_{ij} z_{ij}$ vanishes on $H(f)$. As a corollary of this, *if Y is an apolar scheme to f, then $H(f)$ is contained in the linear span of $v_2(Y)$*, the image of Y under the second Veronese embedding $v_2 : \mathbb{P}(V^*) \to \mathbb{P}(S^2(V^*))$. Indeed: every linear equation $\sum_{i \leq j} a_{ij} z_{ij}$ on $v_2(Y)$ comes from a quadratic equation $\sum_{i \leq j} a_{ij} y_i y_j$ on Y, which by the above also vanishes on $H(f)$.

4.2 Normal forms for cubics

It is possible to classify the cubic surfaces up to linear transformation, and use this
to provide a list of normal forms so that every quaternary cubic can be brought
in one of the normal forms by a linear change of coordinates. This was first done
by Schläfli [Sch63], we refer the interested reader to [Sch97] for an overview.

We first recall the classification of smooth cubic surfaces.

Theorem 4.2.1 (See [Seg42, §§84 - 91]). *Every smooth cubic surface can after
a linear change of coordinates be written in one of the following 4 normal forms:*

1. *Sylvester's pentahedral form:*
$$c_0 x_0^3 + c_1 x_1^3 + c_2 x_2^3 + c_3 x_3^3 + c_4(-x_0 - x_1 - x_2 - x_3)^3 = 0, \qquad (4.2.1)$$
with $c_i \in \mathbb{C}^*$, and $\sum_i \pm \frac{1}{\sqrt{c_i}} \neq 0$.

2. *General rank 6 cubic surfaces:*
$$x_1^3 + x_2^3 + x_3^3 - x_0^2(\mu x_0 + 3\lambda_1 x_1 + 3\lambda_2 x_2 + 3\lambda_3 x_3) = 0 \qquad (4.2.2)$$
with $\lambda_i \in \mathbb{C}^*$, $\mu \in \mathbb{C}$, and $\mu + 2(\lambda_1^{\frac{3}{2}} + \lambda_2^{\frac{3}{2}} + \lambda_3^{\frac{3}{2}}) \neq 0$.

3. *Special rank 6 cubic surfaces:*
$$2\mu_0 x_0^3 + x_1^3 + x_2^3 - 3x_0(\mu_1 x_0 x_1 + x_0 x_2 + x_3^2) = 0, \qquad (4.2.3)$$
with $\mu_1(\mu_0 \pm \mu_1^{\frac{3}{2}} \pm 1) \neq 0$.

4. *Cyclic cubic surfaces:*
$$x_0^3 + x_1^3 + x_2^3 + x_3^3 - 3\lambda x_1 x_2 x_3 = 0, \qquad (4.2.4)$$
with $(\lambda^3 + 8)(\lambda^3 - 1) \neq 0$.

*The families of cubics with these normal forms have respective codimensions
0,1,2,3 in $\mathbb{P}^{19}$. Cubics of the form (4.2.1) have Waring rank 5, cubics of the
form (4.2.2) or (4.2.3) have rank 6, and cubics of the form (4.2.4) have rank 5
if $\lambda \neq 0$, and rank 4 if $\lambda = 0$.*

A detailed discussion on normal forms for singular cubic surfaces can be found
in [BW79]. For our purposes, it suffices to know the following result.

Theorem 4.2.2 (See [BW79, Lemma 2]). *A general singular cubic surface can
be written in the form*
$$x_3(x_1^2 - x_0 x_2) + x_1(x_0 - (1 + \rho_0)x_1 + \rho_0 x_2)(x_0 - (\rho_1 + \rho_2)x_1 + \rho_1 \rho_2 x_2) = 0, \quad (4.2.5)$$
where $\rho_i \in \mathbb{C} \setminus \{0, 1\}$ are pairwise different.

In fact, all we need to know to prove our main theorem is the following.

Corollary 4.2.3. *Every cubic, outside of a certain codimension > 1 set in $\mathbb{P}^{19}$,
can after a linear change of coordinates be written in one of the forms (4.2.1),
(4.2.2) or (4.2.5).*

4.2.1   Sylvester's pentahedral form

Proposition 4.2.4 (See also [Dol12, Chapter 9.4.2]). *No cubic of the form (4.2.1) lies on the Hessian discriminant, as long as all c_i are nonzero.*

Proof. Write $x_4 = -x_0 - x_1 - x_2 - x_3$. The set of rank 2 quadratic forms in $H(f)$ is given by

$$\{c_i x_i^2 - c_j x_j^2 | i \leq j\}.$$

This can easily be verified by hand, or computationally by using the algorithm described below. Since we assumed that all c_i are nonzero, we find that $H(f) \cap X_2$ consists of 10 distinct points, proving the result. $\square$

Remark 4.2.5. We can use our Macaulay2 code to simultaneously analyze $H(f) \cap X_2$ for all cubics f of the form (4.2.1), including the ones where one of the c_i is zero (if two or more of them are zero then $H(f)$ is not defined). The code (available at `https://software.mis.mpg.de/`) computes a primary decomposition of the ideal defining $H(f) \cap X_2$ (where the c_i are variables). The primary decomposition of our ideal has 40 components. 30 of these contain one of the paramaters $c_0, \ldots, c_4$ (each parameter in 6 components); the other 10 do not contain any linear combination of the parameters. This means that if exactly one of the parameters is zero, the intersection $H(f) \cap X_2$ consists of 6 lines, whereas if all the c_i are nonzero, it consists of 10 points. After identifying $H(f)$ with $\mathbb{P}^3$ using the Hessian matrix (as in Sections 4.1.2 and 4.1.3), the 10 points in $H(f)$ are $[1:0:0:0], [0:1:0:0], [0:0:1:0], [0:0:0:1], [1:-1:0:0], [1:0:-1:0], [1:0:0:-1], [0:1:-1:0], [0:1:0:-1], [0:0:1:-1]$.

Remark 4.2.6. Proposition 4.2.4 can also be shown using apolarity: for a general cubic surface

$$f := L_1^3 + L_2^3 + L_3^3 + L_4^3 + L_5^3 = 0$$

(with the L_i in general position) there is an apolar scheme $Y = \{L_1, \ldots, L_5\}$. This can easily be verified directly, but also follows from the so-called *apolarity lemma* [IK99, Lemma 1.15], which states that a homogeneous degree d polynomial f can be written as a linear combination of powers $L_1^d, \ldots, L_s^d$ of linear forms if and only if $\{L_1, \ldots, L_s\}$ is an apolar scheme to f. It now follows from Observation 4.1.9 that $H(f)$ is contained in the linear span $\langle v_2(Y) \rangle$ of the second Veronese embedding of Y.

Clearly, $\langle v_2(Y) \rangle \cap X_2$ contains the 10 lines through the $\langle L_i^2, L_j^2 \rangle$, and (using the fact that L_i are in general position) it is easy to verify that this is in fact an equality. Now, $H(f) \cap X_2$ is the intersection of $\langle v_2(Y) \rangle \cap X_2$ with a hyperplane H. Since $H(f)$ does not contain any of the L_i^2 (indeed: this would imply that there is a g such that $g \circ L_i = 0$ for 4 out 5 of the L_i, contradicting the general position), H intersects every line $\langle L_i^2, L_j^2 \rangle$ in one point, and these points are distinct. These are the 10 points of $H(f) \cap X$.

4.2.2 Rank six cubics

Proposition 4.2.7 (See also [Seg42, §91]). *For a cubic f of the form (4.2.2), the scheme-theoretic intersection $H(f) \cap X_2$ consists of 4 simple and 3 double points. In particular, f lies on the Hessian discriminant.*

Proof. The scheme $H(f) \cap X_2$ is supported at the 7 points $x_1^2 - \lambda_1 x_0^2$, $x_2^2 - \lambda_2 x_0^2$, $x_3^2 - \lambda_3 x_0^2$, $\lambda_1 x_2^2 - \lambda_2 x_1^2$, $\lambda_1 x_3^2 - \lambda_3 x_1^2$, $\lambda_2 x_3^2 - \lambda_3 x_2^2$, $x_0(\mu x_0 + 2\lambda_1 x_1 + 2\lambda_2 x_2 + 2\lambda_3 x_3)$, where the first three are double points. This can be verified using our code. $\square$

Remark 4.2.8. After identifying $H(f)$ with $\mathbb{P}^3$, the 7 points in $H(f)$ are $[0 : 1 : 0 : 0], [0 : 0 : 1 : 0], [0 : 0 : 0 : 1], [0 : \lambda_2 : -\lambda_1 : 0], [0 : \lambda_3 : 0 : -\lambda_1], [0 : 0 : \lambda_2 : -\lambda_3], [1 : 0 : 0 : 0]$, where the first 3 are double points.

Remark 4.2.9. There is an intuitive explanation why $H(f) \cap X_2$ contains 3 double points. As we will see in Section 4.3.1, a cubic of the form (4.2.2) can be obtained as a limit of cubics of the form $\sum_{i=1}^{5} L_i^3$, where in the limit the points $L_4, L_5 \in \mathbb{P}^3$ crash together. In Remark 4.2.6 we saw that for cubics in pentahedral form, the 10 points in $H(f) \cap X_2$ are in bijection with the 10 lines between the 5 points $L_i^2 \in \mathbb{P}^9$. Now if 2 of our points crash together, these 10 lines become 4 simple lines and 3 double lines. We will now make this more precise.

For a general rank 6 cubic surface

$$f := L_1^3 + L_2^3 + L_3^3 + L_4^2 M = 0$$

(with L_1, L_2, L_3, L_4, M in general position) let Z be the nonreduced scheme of length 2 supported at L_4 in direction M, i.e. $I(Z) = I(L_4)^2 + I(\langle L_4, M \rangle)$. Then $Y = L_1 \cup L_2 \cup L_3 \cup Z$ is a length 5 apolar scheme of f. Hence $H(f) \subset \langle v_2(Y) \rangle$.

Note that $v_2(Y) = \langle L_1^2, L_2^2, L_3^2, L_4^2, L_4 M \rangle$. From this we can see that $\langle v_2(Y) \rangle \cap X_2$ contains the 4 lines $\langle L_1^2, L_2^2 \rangle$, $\langle L_1^2, L_3^2 \rangle$, $\langle L_2^2, L_3^2 \rangle$ and $\langle L_4^2, L_4 M \rangle$, as well as three double lines defined by $I(\langle L_i^2, L_4^2 \rangle)^2 + I(\langle L_i^2, L_4^2, L_4 M \rangle)$. As before, using that L_i and M are general we can see that $\langle v_2(Y) \rangle \cap X_2$ consists precisely of these lines. Now, $H(f) \cap X_2$ is the intersection of $\langle v_2(Y) \rangle \cap X_2$ with a hyperplane H. Since $H(f)$ does not contain any of the L_i^2, H intersects every of our 7 lines in 1 (possibly fat) point. These are the 7 points of $H(f) \cap X_2$.

Remark 4.2.10. For cubics of the form (4.2.3), we can use our code to show that $H(f) \cap X_2$ consists of 3 triple points and one single point. After identifying $H(f)$ with $\mathbb{P}^3$, the 4 points in $H(f)$ are $[0 : 1 : 0 : 0], [0 : 0 : 1 : 0], [0 : 0 : 0 : 1], [0 : 1 : -\mu_1 : 0]$, where the first 3 are triple points. In particular, cubics of the form (4.2.3) also lie on the Hessian discriminant, and we recover Observation 4.1.3.

4.2.3 Generic singular cubics

Proposition 4.2.11. *A general singular cubic does not lie on the Hessian discriminant.*

Proof. It suffices to find one singular cubic that does not lie on the Hessian discriminant. One way of doing this is by generating a random one of the form (4.2.5) and using our code. Here we will instead exhibit a very specific example: the Cayley cubic, given by

$$f := x_0 x_1 x_2 + x_0 x_1 x_3 + x_0 x_2 x_3 + x_1 x_2 x_3 = 0,$$

with 4 singular points $[1:0:0:0]$, $[0:1:0:0]$, $[0:0:1:0]$, $[0:0:0:1]$.

Then $H(f)$ is the linear span of the quadratic forms

$$x_1 x_2 + x_1 x_3 + x_2 x_3, x_0 x_2 + x_0 x_3 + x_2 x_3, x_0 x_1 + x_0 x_3 + x_1 x_3, x_0 x_1 + x_0 x_2 + x_1 x_2,$$

and $H(f) \cap X_2$ consists of the following 10 distinct points:

$$x_0(x_1 + x_2 + x_3), x_1(x_0 + x_2 + x_3), x_2(x_0 + x_1 + x_3), x_3(x_0 + x_1 + x_2),$$
$$(x_0 - x_1)(x_2 + x_3), (x_0 - x_2)(x_1 + x_3), (x_0 - x_3)(x_1 + x_2),$$
$$(x_1 - x_2)(x_0 + x_3), (x_1 - x_3)(x_0 + x_2), (x_2 - x_3)(x_0 + x_1).$$

This shows that the f does not lie on the Hessian discriminant. $\qquad \square$

Remark 4.2.12. After identifying $H(f)$ with $\mathbb{P}^3$, the 10 points in $H(f)$ are $[1:1:1:-1], [1:1:-1:1], [1:-1:1:1], [-1:1:1:1], [1:-1:0:0], [1:0:-1:0], [1:0:0:-1], [0:1:-1:0], [0:1:0:-1], [0:0:1:-1]$. It can easily be verified that these 10 points, together with the 4 singular points of f, are precisely the 14 singular points of the Hessian surface of f.

4.2.4   Proof of Proposition 4.1.4

We now have the required background to give a proof of Proposition 4.1.4.

Proof of Proposition 4.1.4. We can verify the proposition separately for the four cases in Theorem 4.2.1. For the first three cases, the singularities of the Hessian surface were computed in [DvG07, §§1.5, 5.2, 5.3]; they agree with the points in $H(f) \cap X_2$ that we obtained in Remarks 4.2.5, 4.2.8 and 4.2.10. For the final case, the Hessian matrix is equal to

$$3 \begin{pmatrix} 2x_0 & 0 & 0 & 0 \\ 0 & 2x_1 & -\lambda x_3 & -\lambda x_2 \\ 0 & -\lambda x_3 & 2x_2 & -\lambda x_1 \\ 0 & -\lambda x_2 & -\lambda x_1 & 2x_3 \end{pmatrix},$$

Hence the Hessian surface $\mathrm{Hess}(f)$ is the union of the plane L defined by $x_0 = 0$ and cone C defined by $(4 - \lambda^3)x_1 x_2 x_3 - \lambda^2(x_1^3 + x_2^3 + x_3^3) = 0$. If $\lambda \neq 0$, one checks (using the assumption $\lambda^3 + 8 \neq 0$) that $\mathrm{Sing}(\mathrm{Hess}(f)) = (L \cap C) \cup \{[1:0:0:0]\}$; if $\lambda = 0$ then $\mathrm{Sing}(\mathrm{Hess}(f))$ is the union of the six lines $x_i = x_j = 0$. In both cases we see that every singular point in the Hessian surface of f gives a rank ≤ 2 matrix, proving our result. $\qquad \square$

4.3 Fundamental invariants

The natural action of SL_4 on V induces an action on the space $S^3(V^*)$ of quaternary cubics, which in turn induces an action on the polynomial ring $R = \mathbb{C}[c_{000}, \ldots, c_{333}]$ in the 20 coefficients of a quaternary cubic. Then the invariant ring R^{SL_4} is the ring of all polynomials in the coefficients of a cubic surface that are invariant under a (determinant 1) linear change of coordinates. It was shown by Salmon [Sal60] that R^{SL_4} is generated by polynomials $I_8, I_{16}, I_{24}, I_{32}, I_{40}, I_{100}$, where I_d has degree d. The first 5 are algebraically independent and I_{100}^2 can be written as a polynomial in $I_8, I_{16}, I_{24}, I_{32}, I_{40}$. Using the connectedness of SL_4, one can show that the fundamental invariants I_d are irreducible. The expressions for I_d in terms of c_{ijk} are hard to obtain and too long to write down here. However, it is easy to write them down for cubics in Sylvester normal form.

For a cubic of the form (4.2.1), we write

$$
\begin{aligned}
\sigma_1 =\ & c_0 + c_1 + c_2 + c_3 + c_4 \\
\sigma_2 =\ & c_0 c_1 + c_0 c_2 + c_0 c_3 + c_0 c_4 + c_1 c_2 + c_1 c_3 + c_1 c_4 + c_2 c_3 + c_2 c_4 + c_3 c_4 \\
\sigma_3 =\ & c_0 c_1 c_2 + c_0 c_1 c_3 + c_0 c_1 c_4 + c_0 c_2 c_3 + c_0 c_2 c_4 \\
& + c_0 c_3 c_4 + c_1 c_2 c_3 + c_1 c_2 c_4 + c_1 c_3 c_4 + c_2 c_3 c_4 \\
\sigma_4 =\ & c_0 c_1 c_2 c_3 + c_0 c_1 c_2 c_4 + c_0 c_1 c_3 c_4 + c_0 c_2 c_3 c_4 + c_1 c_2 c_3 c_4 \\
\sigma_5 =\ & c_0 c_1 c_2 c_3 c_4.
\end{aligned}
$$

Then we can write the fundamental invariants as follows:

$$
\begin{aligned}
I_8 =\ & \sigma_4^2 - 4\sigma_3 \sigma_5 \\
I_{16} =\ & \sigma_5^3 \sigma_1 \\
I_{24} =\ & \sigma_5^4 \sigma_4 \\
I_{32} =\ & \sigma_5^6 \sigma_2 \\
I_{40} =\ & \sigma_5^8.
\end{aligned}
$$

Remark 4.3.1. The tuple $[I_8, I_{16}, I_{24}, I_{32}, I_{40}]$ gives a point in weighted projective space $\mathbb{P}(1, 2, 3, 4, 5)$. We will denote this point by $\mathrm{Inv}(\mathcal{C})$.

4.3.1 Computing invariants for cubics of higher rank

A general cubic of the form (4.2.2) has Waring rank 6 [Sei19], i.e. cannot be written as a sum of 5 cubes. However, since a generic quaternary cubic can be brought in Sylvester normal form, *any* quaternary cubic $\mathcal{C}$ can be arbitrarily closely approximated by cubics in Sylvester normal form.

We do this for cubics of the form (4.2.2). Fix a cubic $\mathcal{C}$ with equation

$$
f(x_0, x_1, x_2, x_3) := x_1^3 + x_2^3 + x_3^3 - x_0^2(\mu x_0 + 3\lambda_1 x_1 + 3\lambda_2 x_2 + 3\lambda_3 x_3) = 0.
$$

For every $\varepsilon \in \mathbb{C}^*$, we define a cubic $\mathcal{C}_\epsilon$ with equation

$$f_\varepsilon(x_0, x_1, x_2, x_3) := \frac{1}{\lambda_1^3 \varepsilon^3}(\varepsilon \lambda_1 x_1)^3 + \frac{1}{\lambda_2^3 \varepsilon^3}(\varepsilon \lambda_2 x_2)^3 + \frac{1}{\lambda_3^3 \varepsilon^3}(\varepsilon \lambda_3 x_3)^3 +$$

$$(\frac{1}{\varepsilon} - \mu)x_0^3 + \frac{1}{\varepsilon}(-x_0 - \varepsilon \lambda_1 x_1 - \varepsilon \lambda_2 x_2 - \varepsilon \lambda_3 x_3)^3 = 0.$$

Note that $\lim_{\varepsilon \to 0} f_\varepsilon = f$. For fixed ε, we can compute $\mathrm{Inv}(\mathcal{C}_\varepsilon) \in \mathbb{P}(1,2,3,4,5)$ (see our code at `https://software.mis.mpg.de/`). Taking the limit $\varepsilon \to 0$ yields

$$\mathrm{Inv}(\mathcal{C}) = \ [\mu^2 - 4(\lambda_0^3 + \lambda_1^3 + \lambda_2^3) : \lambda_0^3\lambda_1^3 + \lambda_0^3\lambda_2^3 + \lambda_1^3\lambda_2^3 : 2\lambda_0^3\lambda_1^3\lambda_2^3 :$$
$$\lambda_0^3\lambda_1^3\lambda_2^3(\lambda_0^3 + \lambda_1^3 + \lambda_2^3) : 0],$$

as was already computed in [DvG07, Theorem 6.6]. In particular, we can deduce the following result (see also [Dol12, Chapter 9.4.5]).

Proposition 4.3.2. *For a general smooth cubic of rank 6, it holds that $I_{40} = 0$.*

4.4 Proof of the main theorem

Theorem 4.4.1. *Let HD be the degree 120 polynomial in $c_{000}, \dots, c_{333}$ obtained by evaluating the Hurwitz form Hu_X of the variety of rank 2 matrices in the Plücker coordinates of $H(f)$, where f defines a general cubic surface. Then $HD = I_{40}^3$, where I_{40} is the degree 40 Salmon invariant.*

Proof. We will show that $V(HD) = V(I_{40})$. Then the result follows from the fact that $\deg(HD) = 3\deg(I_{40})$. The set of cubics that can be brought in the form (4.2.2) is of codimension one (see Theorem 4.2.1), lies on the Hessian discriminant by Proposition 4.2.7, and satisfies $I_{40} = 0$ by Proposition 4.3.2. This, together with irreducibility of I_{40}, implies that $V(HD) \supseteq V(I_{40})$.

Now if this were a strict inclusion, this would mean that $HD = I_{40} \cdot g$, and so $V(HD) = V(I_{40}) \cup V(g)$, with $V(g) \neq V(I_{40})$. Then $V(g)$ is a codimension one set of cubics lying on the Hessian discriminant. But Corollary 4.2.3, and Propositions 4.2.4 and 4.2.11 show that the set of cubics on $V(HD)$ that cannot be brought in the form (4.2.2) is of codimension greater than one, so we reach a contradiction. $\square$

As pointed out in [Øby00], there is an intuitive reason why we would expect HD to be a cube: from Corollary 4.2.3 and Propositions 4.2.4, 4.2.7 and 4.2.11, it follows from that as soon as f lies on the Hessian discriminant, the number of points in $H(f) \cap X_2$ drops from 10 to 7, and not to 9 as expected. Intuitively, this means that f is a triple zero of HD. But then HD is a polynomial for which every root is a triple root, hence it must be a cube.

Part II

Matroids

Chapter 5

Matroids, flag matroids, and homogeneous varieties

In this chapter we review some classical topics on the intersection of matroid theory and geometry. Section 5.1 is an introduction to matroids. After reviewing the most important basic notions in matroid theory, we introduce the Tutte polynomial of a matroid. We also give two less common equivalent definitions of a matroid that will play a role later: via the base polytope and via Gale orderings. In Section 5.2, we present a result of Gelfand, Goresky, MacPherson and Serganova (Theorem 5.2.4), which relates the torus orbits in a Grassmannian to base polytopes of representable matroids.

The second half of this chapter is devoted to generalizing this correspondence between matroids and geometry. On the geometry side, we replace Grassmannians by flag varieties. On the combinatorics side, we replace matroids by flag matroids, which are reviewed in Section 5.3. Then, in Section 5.4, we relate representable flag matroids to flag varieties. This chapter, as well as Chapter 6, are based on the survey paper [CDMS20].

5.1   Matroids

For a comprehensive monograph on matroids we refer the reader to [Oxl11].

5.1.1   Introduction to matroids

There exist many cryptomorphic definitions of a matroid – it can be defined in terms of its independent sets, or its rank function, or its dependent sets, amongst others. One of the most relevant definitions for us is in terms of bases.

Definition 5.1.1. A *matroid* $M = (E, \mathcal{B})$ consists of a ground set E and a collection of subsets $\mathcal{B} \subseteq \mathcal{P}(E)$ such that:

B1. $\mathcal{B} \neq \varnothing$, and

B2. *(basis exchange)* if $B_1, B_2 \in \mathcal{B}$ and $e \in B_1 - B_2$, there exists $f \in B_2 - B_1$ such that $(B_1 - e) \cup f \in \mathcal{B}$.

If $X \subseteq E$ is contained in a basis of M, we say that X is *independent*, and *dependent* otherwise. A minimal dependent set is called a *circuit*. The *rank* $r_M(X)$ of a subset $X \subseteq E$ is defined as the size of the largest independent set contained in X. We will write $r(M)$ for $r_M(E)$. It follows from the matroid axioms that every basis of M has cardinality $r(M)$. We say $X \subseteq E$ is a *flat* if $r(X \cup \{y\}) > r_M(X)$ for all $y \in E \setminus X$.

We can use rank functions to provide an alternative set of axioms to define a matroid. We present this as a lemma, but it can just as well be given as the definition. The diligent reader can check that each set of axioms implies the other.

Lemma 5.1.2. *A matroid $M = (E, r_M)$ can be described by set E and a* rank *function $r_M : \mathcal{P}(E) \to \mathbb{Z}_{\geq 0}$ such that, for $X, Y \in \mathcal{P}(E)$, the following conditions hold:*

R1. $r_M(X) \leq |X|$,

R2. (monotonicity) *if $Y \subseteq X$, then $r_M(Y) \leq r_M(X)$, and*

R3. (submodularity) $r_M(X \cup Y) + r_M(X \cap Y) \leq r_M(X) + r_M(Y)$.

It is also possible to define matroids in terms of their independent sets, circuits, or flats. We refer the reader to the literature for these characterizations.

Example 5.1.3. An easy but fundamental example of a matroid is the *uniform matroid $U_{r,n}$*. It is the matroid on $[n]$ whose bases are given by all subsets of $[n]$ of cardinality r. Its rank function is given by $r_M(X) = \min\{|X|, r\}$.

A reader new to matroid theory should not be surprised by the borrowed terminology from linear algebra: matroids were presented as a generalization of linear independence in vector spaces in the paper by Whitney [Whi35] initiating matroid theory. Matroids also have a lot in common with graphs, thus explaining even more of the terminology used. For instance, very important matroid operations are that of *minors*. These are analogous to the graph operations of the same names. As there, deletion is very simple, while contraction requires a bit more work.

Definition 5.1.4 (Deletion and Contraction).

- We can remove an element e of a matroid $M = (E, r_M)$ by *deleting* it. This yields a matroid $M \backslash e = (E - e, r_{M \backslash e})$, where $r_{M \backslash e}(X) = r_M(X)$ for all $X \subseteq E - e$.

- We can also remove an element e of a matroid $M = (E, r_M)$ by *contracting* it. This gives a matroid $M/e = (E - e, r_{M/e})$ where $r_{M/e}(X) = r_M(X \cup e) - r_M(\{e\})$ for all $X \subseteq E - e$.

Remark 5.1.5. More generally, if $M = (E, r_M)$ is a matroid and S is a subset of E, we can define the deletion $M \backslash S$ (resp. contraction M/S) by deleting (resp. contracting) the elements of S one by one. We have that $r_{M \backslash S}(X) = r_M(X)$ for all $X \subseteq E - S$ and $r_{M/S}(X) = r_M(X \cup S) - r_M(S)$ for all $X \subseteq E - S$.

Definition 5.1.6. Let M_1 and M_2 be 2 matroids on disjoint ground sets E_1 and E_2. Then their *direct sum* $M_1 \oplus M_2$ is the matroid on $E_1 \sqcup E_2$ defined by $\mathcal{B}(M_1 \oplus M_2) = \{B_1 \sqcup B_2 \mid B_1 \in \mathcal{B}(M_1), B_2 \in \mathcal{B}(M_2)\}$. Note that $r_{M_1 \oplus M_2}(X_1 \sqcup X_2) = r_{M_1}(X_1) + r_{M_2}(X_2) = $ for $X_i \subseteq E_i$.

We will now give two examples of classes of matroids which show exactly the relationship matroids have with linear algebra and graph theory. The first one plays a central role in this chapter.

Definition 5.1.7. Let V be a vector space, and $\phi : E \to V$ a map that assigns to every element in E a vector of V. For every subset X of E, define $r(X)$ to be the dimension of the linear span of $\phi(X)$. We have that (E, r) is a matroid, which we say is *representable*. In the literature, representable matroids are sometimes called *realizable* matroids, or *linear* matroids.

Remark 5.1.8. Our definition differs slightly from the one found in literature: typically one identifies E with $\phi(E)$. Our definition does not require ϕ to be injective; we can take the same vector several times. We also note that the matroid represented by $\phi : E \to V$ only depends on the underlying map $\phi : E \to \mathbb{P}(V)$, assuming $\phi(E) \subset V \setminus \{0\}$.

If V is defined over a field $\mathbb{F}$, we say that M is $\mathbb{F}$-*representable*. We can describe the bases of a representable matroid: $X \subseteq E$ is a matroid basis if and only if $\phi(X)$ is a vector space basis of the linear span of $\phi(E)$.

Example 5.1.9 (The non-Pappus matroid). Here is an example of a matroid which is not representable: consider the rank-3 matroid R on $[9]$, whose bases are all 3-element subsets of $[9]$ except for the following:

$$\{1, 2, 3\}, \{4, 5, 6\}, \{1, 5, 7\}, \{1, 6, 8\}, \{2, 4, 7\}, \{2, 6, 9\}, \{3, 4, 8\}, \{3, 5, 9\}.$$

If R were representable over a field $\mathbb{F}$, there would be a map $[9] \to \mathbb{P}^2_{\mathbb{F}} : i \to p_i$ such that p_i, p_j, p_k are collinear if and only if $\{i, j, k\}$ is not a basis of R. Now, the classical *Pappus' Theorem* precisely says that this is impossible: if the non-bases listed above are all collinear, then so are p_7, p_8, p_9.

Definition 5.1.10. Let $G = (V, E)$ be a graph. The *graphic* (or *cycle*) matroid M of G is formed by taking $E(M) = E(G)$, and setting the rank of a set of edges equal to the cardinality of the largest spanning forest contained within it.

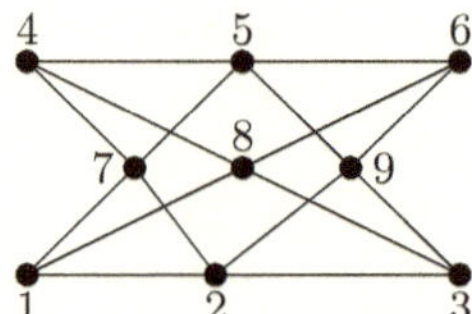

Figure 5.1: The non-Pappus matroid

5.1.2 The Tutte polynomial

Further matroid definitions will be given below, but we have covered enough to introduce the Tutte polynomial, which will play a central role in later chapters. The Tutte polynomial is the most famous matroid (and graph) invariant, and, like matroids themselves, has multiple definitions. Here, we give the *corank-nullity* formula, two terms which will be defined below.

Definition 5.1.11. Let $M = (E, r)$ be a matroid with ground set E and rank function $r : \mathcal{P}(E) \to \mathbb{Z}_{\geq 0}$. The *Tutte polynomial* of M is

$$T_M(x, y) = \sum_{S \subseteq E} (x - 1)^{r(M) - r(S)} (y - 1)^{|S| - r(S)}.$$

The term $r(M) - r(S)$ is called the *corank*, while the term $|S| - r(S)$ is called the *nullity*. Readers familiar with matroid theory should be careful not to confuse a mention of corank with dual rank, given the usual naming convention of dual objects. By identifying the rank function of a matroid with the connectivity function of a graph in an appropriate way, one can pass between this formula and the original formulation of the Tutte polynomial which was given for graphs.

Example 5.1.12. For the (matroid of the) complete graph K_4, there are four subsets with three elements of rank 2 and all the other subsets with three elements have rank 3. In this case, the Tutte polynomial is

$$T_{M(K_4)}(x, y) = x^3 + 3x^2 + 2x + 4xy + 2y + 3y^2 + y^3.$$

Readers interested in seeing what the Tutte polynomial looks like for a range of different classes of matroids should consult [MRIRS12]. The prevalence of the Tutte polynomial in the literature is due to the wide range of applications it has. The simplest of these occurs when we evaluate the polynomial at certain points, these being called *Tutte invariants*. For instance, $T(1, 1)$ gives the number of bases in the matroid (or the number of spanning trees in a graph). In this way we can also count the number of independent sets in a matroid or graph, and the number of acyclic orientations of a graph, as well as some other such quantities.

Beyond numerics, the Tutte invariants also include other well-known polynomials, appearing in graph theory (the chromatic polynomial, concerned with graph colourings; see also Theorem 7.4.6) and network theory (the flow and reliability polynomials). Extending to further disciplines, one can find multivariate versions of the Tutte polynomial which specialize to the Potts model [WM00] from statistical physics and the Jones polynomial [Thi87] from knot theory. In Chapter 6, we will look at the classical Tutte polynomial from an algebraic point of view.

We noted that there are multiple definitions of the Tutte polynomial. One is both so useful and attractive that we would be remiss to not include it. It states that, instead of calculating the full sum above, we can instead simply form a recurrence over minors of our matroid, which can lead to faster calculations. Note that a *coloop* is an element of E which is in every basis of M, while a *loop* is an element which is in no basis.

Lemma 5.1.13 ([BO92])**.** *Let $T_M(x,y)$ be the Tutte polynomial of a matroid $M = (E,r)$. Then the following statements hold.*

    i. $T_M(x,y) = xT_{M/e}(x,y)$ if e is a coloop.

    ii. $T_M(x,y) = yT_{M\setminus e}(x,y)$ if e is a loop.

    iii. $T_M(x,y) = T_{M\setminus e}(x,y) + T_{M/e}(x,y)$ if e is neither a loop nor a coloop.

The Tutte polynomial is in fact universal for such formulae: any formula for matroids (or graphs) involving just deletions and contractions will be an evaluation of the Tutte polynomial. There are numerous proofs of this in the literature, and also extensions to related classes of objects. One such reference is Section 4 of [EMM11]. Note that Lemma 5.1.13 implies that the coefficients of the Tutte polynomial are nonnegative.

We finish this section by listing a couple of easy properties of the Tutte polynomial. Recall that for a matroid M, the *dual matroid* $M^\vee$ is defined by $\mathcal{B}(M^\vee) = \{E \setminus B \mid B \in \mathcal{B}(M)\}$, or equivalently by $r_{M^\vee}(X) = r_M(E \setminus X) + |X| - r(M)$.

Proposition 5.1.14. *Let M be a matroid on E. The following properties hold for T_M:*

1. *(Direct sum) If M is a direct sum $M_1 \oplus M_2$ of two matroids on ground sets E_1, E_2 with $E_1 \sqcup E_2 = E$, then $T_M(x,y) = T_{M_1}(x,y) \cdot T_{M_2}(x,y)$.*

2. *(Loops & coloops) Let ℓ be the number of loops in M, and c the number of coloops in M. Then $x^c y^\ell$ divides $T_M(x,y)$.*

3. *(Duality) If $M^\vee$ is the dual matroid of M, then $T_M(y,x) = T_{M^\vee}(x,y)$.*

Proof. The first and third statement follow immediately from Definition 5.1.11; the second statement follows from Lemma 5.1.13. $\qquad\qquad\square$

5.1.3 The base polytope

We will now give two more axiom systems for matroids. The first one, via base polytopes, will play a fundamental role in this and later chapters. We first define what the base polytope of a matroid is: let $M = (E, \mathcal{B})$ be a matroid. We work in the vector space $\mathbb{R}^E = \{(r_i \mid i \in E)\}$, where $r_i \in \mathbb{R}$. For a set $U \subseteq E$, $e_U \in \mathbb{R}^E$ is the indicator vector of U, that is, e_U is the sum of the unit vectors e_i, for all $i \in U$. Note that $e_{\{i\}} = e_i$.

Definition 5.1.15. The *base polytope* of M is

$$P(M) = \mathrm{conv}\{e_B \mid B \in \mathcal{B}\}.$$

Note that this is always a lattice polytope. We also note that the vertices of $P(M)$ correspond to the bases of M. In particular: given $P(M) \subset \mathbb{R}^E$, we can recover M. The base polytope is a face of the *independent set polytope* of M, which is the convex hull of indicator vectors of the independent sets of M. The dimension of $P(M)$ is determined by the number of *connected components* of M.

Definition 5.1.16. A matroid is *connected* if and only if any two elements are contained in a common circuit. It can be shown that "being contained in a common circuit" is an equivalence relation on E; the equivalence classes are called *connected components*.

Proposition 5.1.17 ([FS05, Proposition 2.4]). *The dimension of $P(M)$ is equal to $|E|$ minus the number of connected components of M.*

Remark 5.1.18. The base polytope $P(M_1 \oplus M_2)$ of a direct sum is equal to the Minkowski sum of $P(M_1) \times \{0\}$ and $\{0\} \times P(M_2)$ in $\mathbb{R}^{E_1} \times \mathbb{R}^{E_2}$.

The following theorem gives a characterization of which lattice polytopes appear as the base polytope of a matroid. It can be used as an axiom system to define matroids. By a $0, 1$-vector, we simply mean a vector all of whose entries are either 0 or 1.

Theorem 5.1.19 ([Edm70], see also [GGMS87, Theorem 4.1]). *A polytope $P \subset \mathbb{R}^E$ is the base polytope of a matroid on E if and only if the following two conditions hold:*

P1. *every vertex of P is a $0, 1$-vector, and*

P2. *every edge of P is parallel to $e_i - e_j$ for some $i, j \in E$.*

In particular, the normal fan of the base polytope $P(M)$ of a matroid is a coarsening of the braid arrangement, which is the normal fan of $\sum_{1 \leq i < j \leq n} \mathrm{Conv}(e_i, e_j)$.

The following proposition is a consequence of the greedy algorithm structure for matroids. For $u \in \mathbb{R}^n$ and a polytope $P \subset \mathbb{R}^n$, let $P^u := \{x \in P \mid \langle x, u \rangle = \max_{y \in P} \langle y, u \rangle\}$ be the face maximizing in the direction of u.

Proposition 5.1.20 ([AK06, Proposition 2]). *Let M be a matroid on E, and let $\mathbf{S} = S_1 \subseteq \cdots \subseteq S_m$ be a flag of subsets of E, and write $e_{\mathbf{S}} = e_{S_1} + \cdots + e_{S_m}$. Then $P(M)^{e_{\mathbf{S}}}$ is the base polytope of the matroid*

$$M_S := M|S_1 \oplus M|S_2/S_1 \oplus \cdots \oplus M|S_m/S_{m-1} \oplus M/S_m.$$

In other words, the bases of M_S are the bases B of M with $\mathrm{rk}_M(S_j) = |B \cap S_j|$ for all $1 \leq j \leq m$.

More information about the faces of matroid base polytopes can be found in [Kim10, FS05]. It follows from Theorem 5.1.19 that the base polytope $P = P(M)$ of a matroid is a *very ample polytope*, meaning that for every vertex v of P, the semigroup generated by the set $\{w - v \mid w \in P \cap \mathbb{Z}^E\}$ is saturated. In fact, an even stronger property holds.

Definition 5.1.21. A lattice polytope P, spanning (as a lattice) the lattice N, is *normal* if and only if for any $k \in \mathbb{Z}_{\geq 0}$ and any $p \in kP \cap N$ we have $p = p_1 + \cdots + p_k$ for some $p_i \in P \cap N$.

Theorem 5.1.22 (White). *For any matroid M the polytope $P(M)$ is normal.*

A proof of White's theorem can be found in [CDMS20, Theorem 3.8].

5.1.4 Definition via Gale orderings

We move on to another axiom system: via Gale orderings. This is orginally due to Gale [Gal68]; our formulation is based on lecture notes by Reiner [Rei05].

Definition 5.1.23. Let ω be a linear ordering on E, which we will denote by $\leq$. Then the *dominance ordering* $\leq_\omega$ on $\binom{E}{k}$, also called *Gale ordering*, is defined as follows. Let $A, B \in \binom{E}{k}$, where

$$A = \{i_1, \ldots, i_k\}, i_1 < \ldots < i_k$$

and

$$B = \{j_1, \ldots, j_k\}, j_1 < \ldots < j_k.$$

Then we set

$$A \leq_\omega B \text{ if and only if } i_1 \leq j_1, \ldots, i_k \leq j_k.$$

Theorem 5.1.24 (Gale, [Gal68]). *Let $\mathcal{B} \subseteq \binom{E}{k}$. We have that $\mathcal{B}$ is the set of bases of a matroid if and only if for every linear ordering ω on E, the collection $\mathcal{B}$ has a maximal element under the Gale ordering $\leq_\omega$ (i.e. there is a unique member $A \in \mathcal{B}$ such that $B \leq_\omega A$, for all $B \in \mathcal{B}$).*

In Section 5.3.1, we will introduce a generalization of matroids, called *flag matroids*, via Gale orderings.

5.2 Grassmannians

As a set, the Grassmannian $Gr(r, V) = Gr(r, n)$, parameterizes r-dimensional subspaces of an n-dimensional vector space V. A point $[L]$ in $Gr(r, n)$ can be represented by a full-rank $r \times n$ matrix A, where our r-dimensional subspace L is the row span of A. Two matrices A and B represent the same point in $Gr(r, n)$ if and only if they are the same up to elementary row operations. $Gr(r, n)$ can be realized as an algebraic variety as follows:

$$Gr(r, n) = Gr(r, V) = \{[v_1 \wedge \ldots \wedge v_r] \subset \mathbb{P}(\overset{r}{\bigwedge} V)\}.$$

Here, $v_1, \ldots, v_r$ are the rows of the aforementioned matrix A, and thus a point of $Gr(r, n)$ is identified with the subspace $L = \langle v_1, \ldots, v_r \rangle$. The embedding presented above is known as the *Plücker embedding* and the Grassmannian is defined by quadratic polynomials known as *Plücker relations* [Man01]. Explicitly, in coordinates, the map associates to the matrix A the value of all $r \times r$ minors.

The Plücker embedding may be identified with a very ample line bundle on $Gr(r, n)$, which we will denote by $\mathcal{O}(1)$. Other very ample line bundles on $Gr(r, n)$ are the d-th tensor powers $\mathcal{O}(d)$. They can be realized as a composition of the Plücker embedding with the d-th Veronese map $\mathbb{P}(\bigwedge^r V) \to \mathbb{P}(\mathrm{Sym}^d \bigwedge^r V)$.

Remark 5.2.1. A reader not familiar at all with very ample line bundles may think about them as maps into projective spaces. Let us present this with the example of the projective space $\mathbb{P}^n$ (which also equals $Gr(1, n + 1)$). We have an identity map $\mathbb{P}^n \to \mathbb{P}^n$, which corresponds to $\mathcal{O}(1)$. The d-th *Veronese map* embeds $\mathbb{P}^n$ in a larger projective space $\mathbb{P}^{\binom{n+d}{n}-1}$ by evaluating on a point all degree d monomials. The associated map is given by $\mathcal{O}(d)$. For $n = 1$ and $d = 2$ we get:

$$\mathbb{P}^1 \ni [x : y] \to [x^2 : xy : y^2] \in \mathbb{P}^2.$$

It will follow from Theorem 5.4.2 that the embedding of $Gr(r, n)$ by $\mathcal{O}(d)$ spans a projectivization of an irreducible representation V_{λ_0} of GL_n. The Young diagram $\lambda_0 = (d, \ldots, d)$ consists of r rows of length d.

5.2.1 Representable matroids and geometry

Let us consider a representable matroid M given by $n = |E|$ vectors spanning an r-dimensional vector space V. By fixing a basis of V we may represent this matroid as an $r \times n$ matrix A. On the other hand the matrix A may be regarded as defining an r-dimensional subspace L of an n-dimensional vector space, i.e. a point in $Gr(r, n)$. Since applying elementary row operations to A does not change which of the maximal minors of A vanish, the matroid M only depends on the r-dimensional subspace, and not on the specific matrix A representing our subspace.

In this way we have associated to any point $p = [L] \in Gr(r, n)$ a representable rank r matroid on $[n]$, which we will denote by $M(L)$ or by M_p. More invariantly, $M(L)$ is the representable matroid whose ground set is the image of $\{e_1, \ldots, e_n\}$ under the dual map $\mathbb{C}^n \twoheadrightarrow L^\vee$.

The vector space $\mathbb{C}^n$ comes with the action of a torus $T = (\mathbb{C}^*)^n$, which induces a T-action on $Gr(r, \mathbb{C}^n)$. We have associated a point $p \in Gr(r, \mathbb{C}^n)$ to a representation of a matroid. If we change the representation by rescaling the vectors we do not change the matroid and the associated point belongs to the orbit Tp. Hence, the intrinsic properties of the matroid M_p should be related to the geometry of Tp – a feature we will examine in detail. The closure $\overline{Tp}$ is a projective toric variety. For more information about toric geometry we refer to [CLS11, Stu96, Ful93, Mic18].

Remark 5.2.2. Of course it can happen that different torus orbits give rise to the same matroid: there are only finitely many matroids on $[n]$, but if $1 < r < n-1$ there are infinitely many torus orbits in $Gr(r, n)$. In fact, the set of all points in $Gr(r, n)$ giving rise to the same matroid forms a so-called *thin Schubert cell* or *matroid stratum*, which typically is a union of infinitely many torus orbits. Thin Schubert cells were first introduced in [GGMS87]. Thin Schubert cells are badly behaved in general: for fixed $r \geq 3$ the thin Schubert cells of $Gr(r, n)$ exhibit arbitrary singularities if n is large enough. This is a consequence of Mnëv's theorem [Mnë88]. See [Laf03, Section 1.8] for a more detailed discussion.

Before stating the promised correspondence between the matroid M_p and the variety $\overline{Tp}$, we need to recall one definition from toric geometry.

Definition 5.2.3. [CLS11, Chapter 2] Let $P \subseteq \mathbb{R}^n$ be a very ample lattice polytope, and write $P \cap \mathbb{Z}^n = p_1, \ldots, p_s$. The *toric variety associated to P* is the Zariski closure of the image of the map $T = (\mathbb{C}^*)^n \to \mathbb{P}^{s-1}$ given by

$$\mathbf{t} \mapsto [\mathbf{t}^{p_1} : \ldots : \mathbf{t}^{p_s}].$$

Theorem 5.2.4 (Gelfand-Goresky-MacPherson-Serganova [GGMS87]). *For $p \in Gr(r, n)$, the projective toric variety associated to $P(M_p)$ is isomorphic to the torus orbit closure $\overline{Tp}$.*

Proof. Let A be a matrix whose rows span the space corresponding to p. The parameterization of $\overline{Tp}$ is given by:

$$\phi : T \to \mathbb{P}(\overset{r}{\bigwedge} \mathbb{C}^n).$$

The coordinates of the ambient space are indexed by r-element subsets of the n columns of the matrix A. The Plücker coordinate indexed by I of $\phi(t_1, \ldots, t_n)$

equals $\prod_{i \in I} t_i$ times the $r \times r$ minor of A determined by I, which we will denote by $\det(A_I)$. In other words, the map ϕ in Plücker coordinates is given as follows:

$$\phi(t_1, \ldots, t_n) = (\det(A_I) \cdot \prod_{i \in I} t_i)_{I \in \binom{[n]}{r}}.$$

The I-th coordinate is nonzero if and only if I is a basis of M_p. Hence, the ambient space of $\overline{Tp}$ has coordinates indexed by basis elements of M_p. After restricting to this ambient space and composing with the isomorphism inverting the nonzero minors $\det(A_I)$, our map can be written as

$$\phi(t_1, \ldots, t_n) = (\prod_{i \in I} t_i)_{e_I \in P(M_p)}.$$

This is exactly the construction of the toric variety represented by $P(M_p)$. $\qquad\square$

It is a major problem to provide the algebraic equations of $\overline{Tp}$. This is equivalent to finding integral relations among the basis of a matroid. We point out that matroids satisfy a 'stronger' property then one could expect from the basis exchange axiom B2 of Definition 5.1.1. Precisely, for any two bases $B_1, B_2 \in \mathcal{B}$ and a subset $A \subset B_1 - B_2$, there exists $A' \subset B_2 - B_1$ such that $(B_1 - A) \cup A'$ and $(B_2 - A') \cup A$ are in $\mathcal{B}$ [Gre73]. This exactly translates to a binomial quadric (degree 2 polynomial) in the ideal of $\overline{Tp}$: $x_{B_1} x_{B_2} - x_{(B_1-A) \cup A'} x_{(B_2-A') \cup A}$, where, as in the proof of Theorem 5.2.4, we label each coordinate by a basis of the matroid. Further, if $|A| = 1$ we obtain special quadrics corresponding to exchanging one element in a pair of bases. The following conjecture due to White provides a full set of generators for any matroid M.

Conjecture 5.2.5. *The ideal of the toric variety represented by $P(M)$ is generated by the special quadrics corresponding to exchanging one element in a pair of bases.*

We note that it is unknown whether the ideal of this toric variety is generated by quadrics, or that all quadrics are spanned by the special quadrics described above. However, it is known that the special quadrics define the variety as a set (or more precisely as a projective scheme) [LM14, Las16].

Combinatorial methods can be used to prove geometric properties of torus orbit closures in Grassmannians. For instance, normality of a polytope corresponds to *projective normality* of the associated toric variety [CLS11, Chapter 2], [Stu96] (less formally, the associated toric variety is not very singular and is embedded in a particularly nice way in the projective space). Thus, White's theorem 5.1.22 has the following geometric consequence.

Corollary 5.2.6. *Any torus orbit closure in any Grassmannian is projectively normal.*

5.3 Flag matroids

Flag matroids first arose as a special case of the so-called *Coxeter matroids*, introduced by Gelfand and Serganova [GS87b, GS87a]. In this section we give a combinatorial introduction to flag matroids. The exposition is largely based on Chapter 1 of [BGW03].

5.3.1 Definition

We start by defining flag matroids in the way they are usually defined in the literature: using Gale orderings.

Definition 5.3.1. Let $0 < r_1 \leq \ldots \leq r_k < n$ be natural numbers. Let $\mathbf{r} = (r_1, \ldots, r_k)$. A *flag F of rank $\mathbf{r}$ on E* is an increasing sequence

$$F_1 \subseteq F_2 \subseteq \cdots \subseteq F_k$$

of subsets of E such that $|F_i| = r_i$ for all i. The set of all such flags is $\mathcal{F}_E^{\mathbf{r}}$.

Let ω be a linear ordering on E. We can extend the Gale ordering $\leq_\omega$ to flags:

$$(F_1, \ldots, F_k) \leq_\omega (G_1, \ldots, G_k) \text{ if and only if } F_i \leq_\omega G_i \text{ for all } i.$$

Definition 5.3.2. A *flag matroid $\mathcal{F}$ of rank $\mathbf{r}$ on E* is determined by a collection $\mathcal{B}(\mathcal{F})$ of flags in $\mathcal{F}_E^{\mathbf{r}}$, which we call *bases*, satisfying the following property: for every linear ordering ω on E, the collection $\mathcal{B}(\mathcal{F})$ contains a unique element which is maximal in $\mathcal{B}(\mathcal{F})$ with respect to the Gale ordering $\leq_\omega$.

If $\mathcal{F}$ is a flag matroid, the collection $\{F_i \mid F \in \mathcal{B}(\mathcal{F})\}$ is called the *i-th constituent* of $\mathcal{F}$. This is clearly the set of bases of a matroid (of rank r_i).

Remark 5.3.3. In the literature it is usually required that we have strict inequalities $0 < r_1 < \ldots < r_k < n$. From a combinatorial point of view this does not make a difference, but when we later consider flag matroid polytopes this restriction would appear artificial. This is also the reason why in Section 5.4 we will not just consider flag varieties, but also their Veronese re-embeddings.

5.3.2 Matroid quotients and representable flag matroids

Next, we want to describe which tuples of matroids can arise as the constituents of a flag matroid. In order to give this characterization, we first need to recall *matroid quotients.*

Definition 5.3.4. Let M_1 and M_2 be matroids on the same ground set E. We say that M_1 is a *quotient* of M_2, written $M_1 \twoheadleftarrow M_2$ if one of the following equivalent statements holds:

(i) Every flat of M_1 is a flat of M_2.

(ii) Every circuit of M_2 is a union of circuits of M_1.

(iii) If $X \subseteq Y \subseteq E$, then $r_{M_2}(Y) - r_{M_2}(X) \geq r_{M_1}(Y) - r_{M_1}(X)$.

(iv) There exists a matroid R and a subset X of $E(R)$ such that $M_2 = R \backslash X$ and $M_1 = R/X$.

(v) For all bases B of M_2 and all $x \notin B$, there is a basis B' of M_1 with $B' \subseteq B$ and such that $\{y : (B' - y) \cup x \in \mathcal{B}(M_1)\} \subseteq \{y : (B - y) \cup x \in \mathcal{B}(M_2)\}$.

For the equivalence of (i), (ii), (iii) and (iv), we refer to [Oxl11, Proposition 7.3.6]. Part (v) is left to the reader. Here are some basic properties of matroid quotients.

Proposition 5.3.5. *Let M_1 be a quotient of M_2.*

(i) *Every basis of M_1 is contained in a basis of M_2, and every basis of M_2 contains a basis of M_1.*

(ii) $r(M_1) \leq r(M_2)$ *and in case of equality* $M_1 = M_2$.

Proof. Both statements can be easily deduced by plugging in $Y = E$ or $X = \emptyset$ in Definition 5.3.4 (iii). $\qquad\square$

A matroid quotient $M_1 \twoheadleftarrow M_2$ is an *elementary quotient* if $r(M_2) - r(M_1) = 1$. Every matroid quotient $M_1 \twoheadleftarrow M_2$ can be realized as a composition of a series of elementary quotients. A canonical one is given by the *Higgs factorization*

$$M_1 = M^{(r(M_2)-r(M_1))} \twoheadleftarrow \cdots \twoheadleftarrow M^{(1)} \twoheadleftarrow M^{(0)} = M_2,$$

which is defined by

$$\mathcal{B}(M^{(i)}) = \{S \subseteq E \mid |S| = r(M_2) - i,\ S \text{ spans } M_1 \text{ and is independent in } M_2\}.$$

The subsets $S \subseteq E$ that span M_1 and are independent in M_2 are called *pseudobases* of (M_1, M_2). We now come to the promised characterization of constituents of flag matroids. In fact, it will turn out we can use it as an alternative definition of flag matroids.

Definition 5.3.6. We call a tuple $(M_1, \ldots, M_k)$ of matroids *concordant* if for every pair (M_i, M_j) with $i < j$, M_i is a quotient of M_j. Since "being a quotient of" is transitive, this is equivalent to requiring M_i to be a quotient of M_{i+1} for $i = 1, \ldots, k-1$.

Theorem 5.3.7 ([BGW03, Theorem 1.7.1]). *A collection $\mathcal{B}$ of flags in $\mathcal{F}_E^{\mathbf{r}}$ is a flag matroid if and only if the following three conditions hold:*

1. *Every constituent $M_i := \{F_i \mid F \in \mathcal{B}\}$ is a matroid.*

2. *The matroids $M_1, \ldots, M_k$ are concordant.*

3. *Every flag $B_1 \subseteq \ldots \subseteq B_k$, with B_i a basis of M_i, is in $\mathcal{B}$.*

In other words, flag matroids on E are in one-to-one correspondence with tuples of concordant matroids on E. If $\mathcal{F}$ is the unique flag matroid with constituents $M_1, \ldots, M_k$, we will often write $\mathcal{F} = (M_1, \ldots, M_k)$. The next result will be essential for defining representable flag matroids. It also explains where the term "matroid quotient" comes from – below we think of W as a vector space quotient of V.

Proposition 5.3.8 ([Bry86, Proposition 7.4.8 (2)]). *Let V and W be vector spaces and $\phi : E \to V$ be a map. Furthermore, let $f : V \to W$ be a linear map. Consider the matroid M_2 represented by ϕ, and the matroid M_1 represented by $f \circ \phi$. Then M_1 is a matroid quotient of M_2.*

Example 5.3.9. If R is a representable matroid on E and X is a subset of E, then $M_2 := R \backslash X$ and $M_1 := R/X$ are representable matroids, and there is a linear map as in Proposition 5.3.8. Indeed, if R is represented by $\phi : E \to V$, then consider the projection $\pi : V \to V/\langle \phi(X) \rangle$. It is not hard to see that M_2 is represented by $\phi|_{E-X}$ and that M_1 is represented by $\pi \circ \phi|_{E-X}$.

Example 5.3.10. The converse of Proposition 5.3.8 is false: we give an example (taken from [BGW03, Section 1.7.5]) of two representable matroids M_2 and M_1 such that M_1 is a quotient of M_2, but there is no map as in Proposition 5.3.8. Let M_2 be the rank-3 matroid on $[8]$ represented by the following matrix

$$\begin{pmatrix} 1 & 0 & 1 & 0 & 1 & 1 & 0 & 1 \\ 0 & 1 & 1 & 0 & 2 & 2 & 2 & 1 \\ 0 & 0 & 0 & 1 & 1 & 2 & 1 & 1 \end{pmatrix}$$

and let M_1 be the rank-2 matroid on $[8]$ whose bases are all 2-element subsets except for $\{2, 6\}$ and $\{3, 5\}$. It is easy to see that M_1 is a representable matroid: just pick six pairwise independent vectors in the plane, and map 2 and 6, as well as 3 and 5, to the same vector. Now M_1 is a matroid quotient of M_2, since the non-Pappus matroid R from Example 5.1.9 satisfies $M_2 = R \backslash 9$ and $M_1 = R/9$. However, it is not possible to find representations V (resp. W) of M_2 (resp. M_1) such that there is a map $f : V \to W$ as in Proposition 5.3.8. Roughly speaking, the problem is that the "big" matroid R is not representable. For a more precise argument, see [BGW03, Section 1.7.5].

We can now define representable flag matroids:

Definition 5.3.11. Let $V_k \to V_{k-1} \to \cdots \to V_1$ be a sequence of linear maps, let $\phi : E \to V_k$ be a map, and let for every $1 \leq i \leq k$, M_i be the matroid represented by the composition $E \to V_k \to \cdots \to V_i$. By Proposition 5.3.8, the matroids $M_1, \ldots, M_k$ are concordant. The *representable flag matroid* $\mathcal{F}(E \to V_k \to \cdots \to V_1)$ is defined as the unique flag matroid whose constituents are $M_1, \ldots, M_k$.

Remark 5.3.12. Example 5.3.10 shows that it can happen that all constituents of a flag matroid are representable matroids, but still the flag matroid is not representable (because the matroid representations are "not compatible").

5.3.3 Flag matroid polytopes

Given a flag $F = (F_1, \ldots, F_k)$ on $[n]$, we will write $e_F := e_{F_1} + \ldots + e_{F_k}$. Note that if F and F' are flags of the same rank $\mathbf{r}$, then e_F and $e_{F'}$ are equal up to an $\mathfrak{S}_n$-permutation.

Definition 5.3.13. The *base polytope* $P(\mathcal{F})$ of a flag matroid $\mathcal{F}$ on $[n]$ is the convex hull of the set $\{e_F \mid F \in \mathcal{B}(\mathcal{F})\} \subset \mathbb{R}^n$.

Example 5.3.14. Let $\mathcal{F}$ be the rank $(1, 2)$ flag matroid on $[3]$ whose bases are $1 \subseteq 12$, $1 \subseteq 13$, $2 \subseteq 12$ and $3 \subseteq 13$. Then its base polytope is the convex hull of the points $(2, 1, 0), (2, 0, 1), (1, 2, 0), (1, 0, 2)$. Its constituents are the uniform rank 1 matroid on $[3]$, and the rank 2 matroid with bases 12 and 13. The base polytope of this flag matroid is depicted in Figure 5.2. $\mathcal{F}$ is a representable flag matroid: consider for example $\{1, 2, 3\} \to \mathbb{C}^2 \twoheadrightarrow \mathbb{C}$, where the first map is given by $1 \mapsto e_1, 2 \mapsto e_2, 3 \mapsto e_2$, and the second map by $e_i \mapsto 1$.

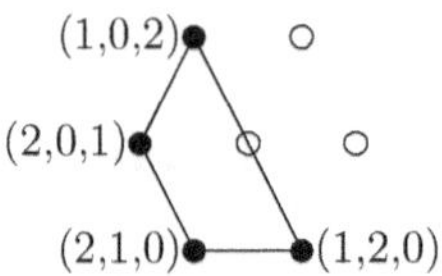

Figure 5.2: A flag matroid base polytope

Theorem 5.3.15 ([BGW03, Theorem 1.11.1]). *A lattice polytope $P \subset \mathbb{R}^n$ is the base polytope of a rank $\mathbf{r}$ flag matroid on $[n]$ if and only if the following two conditions hold:*

1. Every vertex of P is an $\mathfrak{S}_n$-permutation of $e_{\{1,2,\ldots,r_1\}} + \cdots + e_{\{1,2,\ldots,r_k\}}$, and

2. Every edge of P parallel to $e_i - e_j$ for some $i, j \in [n]$.

As for matroids, it follows that the normal fan of the base polytope $P(\mathcal{F})$ of a flag matroid is a coarsening of the braid arrangement.

Theorem 5.3.16 ([BGW03, Corollary 1.13.5.]). *The base polytope of a flag matroid is the Minkowski sum of the matroid base polytopes of its constituents.*

Proposition 5.3.17. *Let $\mathcal{F} = (M_1, \ldots, M_k)$ be a flag matroid on E or rank $\mathbf{r} = (r_1, \ldots, r_k)$, and let $\mathbf{S} = S_1 \subseteq \cdots \subseteq S_m$ be a flag of subsets of E. Then the face $P(\mathcal{F})^{e_{\mathbf{S}}}$ of $P(\mathcal{F})$ maximizing the $e_{\mathbf{S}}$-direction is the base polytope of a flag matroid whose i-th constituent (for $i = 1, \ldots, k$) is*

$$M_i|S_1 \oplus M_i|S_2/S_1 \oplus \cdots \oplus M_i|S_m/S_{m-1} \oplus M_i/S_m.$$

In other words, the bases of the flag matroid of $P(\mathcal{F})^{e_{\mathbf{S}}}$ are bases $F = (B_1, \ldots, B_k)$ of $\mathcal{F}$ such that $\mathrm{rk}_{M_i}(S_j) = |B_i \cap S_j|$ for all $1 \leq i \leq k$ and $1 \leq j \leq m$.

Proof. Note that if $P = \sum_{i=1}^{k} P_i$ is a Minkowski sum of polytopes, then for any $u \in \mathbb{R}^n$, the face P^u is the Minkowski sum $\sum_{i=1}^{k} P_i^u$ of faces. The proof of the proposition is thus reduced to the case of $\mathcal{F}$ being a matroid M, which is Proposition 5.1.20. $\qquad\square$

5.4 Flag varieties

Let us fix a sequence of k positive integers $0 < r_1 \leq \ldots \leq r_k < n$ and let V be an n-dimensional vector space. The (partial) flag variety is defined as follows:

$$Fl(r_1, \ldots, r_k; V) = \{L_1 \subseteq \ldots \subseteq L_k \subset V : \dim L_i = r_i\} \subset \prod_{i=1}^{k} Gr(r_i, V)$$

From now on we will abbreviate the tuple $(r_1, \ldots, r_k)$ to $\mathbf{r}$, and the flag variety $Fl(r_1, \ldots, r_k; n)$ to $Fl(\mathbf{r}, n)$. A point in $Fl(\mathbf{r}, n)$ can be represented by a full-rank $n \times n$-matrix A: the row span of the first r_i rows is L_i. (Although note that only the first r_k rows of the matrix are relevant.) As with Grassmannians, different matrices can represent the same point in $Fl(\mathbf{r}, n)$. More precisely, if we partition the rows of A into blocks of size $r_1, r_2 - r_1, \ldots, n - r_k$, then we are allowed to do row operations on A, with the restriction that to a certain row we can only add a multiple of a row in the same block or a block above. Another way to think about this is the following: let $P_{\mathbf{r}} \subset GL_n(\mathbb{C})$ be the parabolic subgroup of all invertible matrices A with $A_{ij} = 0$ if $i \leq r_s < j$, for some s. Then two $n \times n$ matrices represent the same flag if and only if they are the same up to left multiplication with an element of $P_{\mathbf{r}}$. Hence $Fl(\mathbf{r}, n)$ can also be described as the quotient $_{P_{\mathbf{r}}}\backslash^{GL_n(\mathbb{C})}$ (a homogeneous variety).

The variety $Fl(\mathbf{r}, V)$ comes with a natural embedding

$$Fl(\mathbf{r}, V) \subset \prod_i Gr(r_i, V) \subset \prod_i \mathbb{P}\left(\bigwedge^{r_i} V\right) \subset \mathbb{P}\left(\bigotimes_i \bigwedge^{r_i} V\right), \qquad (5.4.1)$$

where the last map is the Segre embedding. The corresponding line bundle is denoted $\mathcal{O}(1) = \mathcal{O}(1, \ldots, 1)$ (where the number of 1's is equal to k).

Remark 5.4.1. In the literature it is typically assumed that $r_1 < \ldots < r_k$. Indeed, if $\mathbf{r}$ is obtained from $\mathbf{r}'$ by repeating some of the entries, the abstract flag varieties $Fl(\mathbf{r}, V)$ and $Fl(\mathbf{r}', V)$ are isomorphic. However, in this document $Fl(\mathbf{r}, V)$ refers to the flag variety *together with the embedding (5.4.1)*, and this embedding does change if we repeat some entries of $\mathbf{r}'$. Writing $f_{\mathbf{r}}(s)$ for the number of entries in $\mathbf{r}$ equal to s, we note that the image of (5.4.1) is contained in

$$\mathbb{P}\left(\bigotimes_s S^{f_{\mathbf{r}}(s)}\left(\overset{s}{\bigwedge} V\right)\right) \subset \mathbb{P}\left(\bigotimes_i \overset{r_i}{\bigwedge} V\right).$$

Hence, one can think of $Fl(\mathbf{r}, V)$ as a Segre-Veronese reembedding of the flag variety $Fl(\mathbf{r}', V)$. The line bundle on $Fl(\mathbf{r}', V)$ that corresponds to this embedding is $\mathcal{O}(f_{\mathbf{r}}(r_1'), \ldots, f_{\mathbf{r}}(r_k'))$.

5.4.1   Aside: the Borel-Weil theorem

The following beautiful theorem by Borel and Weil relates the embedding (5.4.1) to the representation theory of GL_n. See Section 3.2.1 for a brief introduction to the representation theory of GL_n.

Theorem 5.4.2 (Borel-Weil). *Let* $\mathbf{r} = (r_1, \ldots, r_k)$ *with* $0 < r_1 \leq \ldots \leq r_k < n$, *and let* λ *be the partition conjugate to* $[r_k, \ldots, r_1]$. *Then the linear span of* $Fl(\mathbf{r}; n)$ *under the embedding (5.4.1) is the projectivization of the irreducible* $GL(V)$-*representation* $\mathbb{S}^\lambda V$.

Proof. Fix a basis $e_1, \ldots, e_n$ of V. Let us consider the flag of subspaces

$$\langle e_1, \ldots, e_{r_1} \rangle \subset \langle e_1, \ldots, e_{r_2} \rangle \subset \ldots \subset \langle e_1, \ldots, e_{r_k} \rangle$$

and the corresponding point $p \in Fl(\mathbf{r}, n)$. Under the embedding (5.4.1) it is mapped to

$$(e_1)^{\circ f_{\mathbf{r}}(1)} \otimes (e_1 \wedge e_2)^{\circ f_{\mathbf{r}}(2)} \otimes \ldots \otimes (e_1 \wedge \ldots \wedge e_n)^{\circ f_{\mathbf{r}}(n)} \in S^{f_{\mathbf{r}}(1)}\left(\overset{1}{\bigwedge} V\right) \otimes \ldots \otimes S^{f_{\mathbf{r}}(n)}\left(\overset{n}{\bigwedge} V\right).$$

The $GL(V)$-decomposition of the ambient space is highly non-trivial. However, looking directly at the T-decomposition we see that, up to scaling, the image of p is the unique lexicographically-highest vector. Hence, in particular, the image of p belongs to $\mathbb{S}^\lambda V$, as all other $GL(V)$-representations appearing in the decomposition have strictly smaller highest weights. Furthermore, the flag variety is an orbit under the $GL(V)$-action – one can explicitly write a matrix mapping any flag to any other given flag. Thus, if one point is contained in the irreducible representation, the whole variety must be contained in it.

It remains to show that the span of the flag variety is indeed the whole irreducible representation. This is true, as the flag variety is $GL(V)$-invariant, and thus its linear span is a representation of $GL(V)$. As $\mathbb{S}^\lambda V$ is irreducible, the linear span must coincide with it. $\square$

The above theorem may be regarded as a realization of irreducible $GL(V)$-representations as spaces of sections of a very ample line bundle on a flag variety. A more general Borel-Weil-Bott theorem provides not only a description of global sections – zeroth cohomology – but also higher, arbitrary cohomology.

5.4.2 Representable flag matroids and geometry

Let $\mathbf{L} = (L_1 \subseteq \cdots \subseteq L_k)$ be a flag of subspaces of $\mathbb{C}^n$. Then the dual maps $L_k^\vee \twoheadrightarrow \cdots \twoheadrightarrow L_1^\vee$, together with $\{e_1, \ldots, e_n\} \to \mathbb{C}^n \twoheadrightarrow L_k^\vee$, define a representable flag matroid, which we will denote be $\mathcal{F}(\mathbf{L})$. In coordinates: if F is given by an $n \times n$ matrix A (i.e. the first r_i rows of A span L_i), then $\mathcal{F}(\mathbf{L})$ is the flag matroid whose i-th constituent is the representable matroid corresponding to the submatrix of A consisting of the top r_i rows.

Now, consider the flag variety $Fl(\mathbf{r}, n)$, as described in Section 5.4. The action of the torus $T = (\mathbb{C}^*)^n$ on $\mathbb{C}^n$ induces an action of T on $Fl(\mathbf{r}, n)$. A point $p = \mathbf{L} \in Fl(\mathbf{r}, n)$ gives rise to a representable flag matroid $\mathcal{F}_p := \mathcal{F}(\mathbf{L})$ on $[n]$. All points in the orbit Tp give rise to the same flag matroid. This last statement follows easily from the analogous fact for matroids and the fact that a flag matroid is determined by its constituent matroids. The analogue of Theorem 5.2.4 holds. This was already known by Gelfand, Goresky, MacPherson and Serganova.

Theorem 5.4.3. *For $p \in Fl(\mathbf{r}, n)$, the projective toric variety associated to $P(\mathcal{F}_p)$ is isomorphic to the torus orbit closure $\overline{Tp}$.*

Proof. The proof is a straightforward generalization of the proof of Theorem 5.2.4, with the parameterization of $\overline{Tp}$ given by:

$$\phi : T \to \mathbb{P}\left(\overset{r_1}{\bigwedge} \mathbb{C}^n\right) \times \cdots \times \mathbb{P}\left(\overset{r_k}{\bigwedge} \mathbb{C}^n\right).$$

$\square$

Chapter 6

Equivariant K-theory and the Tutte polynomial

In the previous chapter, we have presented a correspondence between representable matroids and torus orbits in Grassmannians, and generalizations of this correspondence to representable flag matroids. We would like to drop the word "representable" from all of those. As we will see, one way to do this is by replacing "torus orbits" with "classes in equivariant K-theory". This was done for matroids by Fink and Speyer [FS12]. In this chapter, we review their construction, and introduce a generalization to flag matroids.

In Section 6.1, we give an introduction to non-equivariant and equivariant K-theory, including the method of equivariant localization. In Section 6.2 we compute the class of a representable matroid in the K-theory of the Grassmannian, and define the K-class of a general matroid. We generalize this to flag varieties to define the K-class of a flag matroid. Finally, we review the geometric description of the Tutte polynomial of a matroid in terms of its K-class, and use the same construction to define a new invariant: the flag-geometric Tutte polynomial of a flag matroid.

6.1 Equivariant K-theory

6.1.1 Introduction to K-theory and equivariant K-theory

This section is based on [Ful98, Section 15.1]. Let X be an algebraic variety. We define $K^0(X)$ to be the free abelian group generated by vector bundles on X, subject to relations $[E] = [E'] + [E'']$ whenever E' is a subbundle of E, with quotient bundle $E'' = E/E'$. The group $K^0(X)$ inherits a ring structure from the tensor product: $[E] \cdot [F] = [E \otimes F]$. Similarily, we can define $K_0(X)$ to be the free abelian group generated by isomorphism classes of coherent sheaves on X, subject to relations $[A] + [C] = [B]$ whenever there is a short exact sequence

$0 \to A \to B \to C \to 0$. There is an inclusion $K^0(X) \hookrightarrow K_0(X)$. From now on, we will always assume that X is a smooth variety. In this case, the inclusion is an isomorphism, allowing us to identify $K^0(X)$ and $K_0(X)$.

Let $f : X \to Y$ be a map of (smooth) varieties. Then there is a pullback map $f^* : K^0(Y) \to K^0(X)$ defined by $f^*[E] = [f^*E]$ (where E is a vector bundle on Y). If f is a proper map, there is also a pushforward map $f_* : K_0(X) \to K_0(Y)$ given by $f_*[A] = \sum (-1)^i [R^i f_* A]$. Here $R^i f_*$ are right derived functors of the pushforward. An interested reader is advised to find the details in [Ful98, Section 15]. In practice, we will not be using the formal definitions of $K^0(X)$, f^* or f_*. Instead, we will refer to explicit descriptions of those in the cases that we need, each time providing a theorem we build upon.

Remark 6.1.1. In all the cases we study the ring $K^0(X)$ is isomorphic to the cohomology ring and to the Chow ring (after tensoring with $\mathbb{Q}$). Note however that the map from $K^0(X)$ to the Chow ring is nontrivial and given by the Chern character.

Example 6.1.2. Consider the projective space $\mathbb{P}^n$. The (rational) Chow ring is $A(\mathbb{P}^n) = \mathbb{Q}[H]/(H^{n+1})$. Here one should think about H as a hyperplane in $\mathbb{P}^n$ and H^k as a codimension k projective subspace. The most important line bundle is $\mathcal{O}(1)$. The Chern character $ch : K^0(\mathbb{P}^n) \to A(\mathbb{P}^n)$ sends $[\mathcal{O}(1)]$ to $\sum_{i=0}^{n} H^i/i!$. Note that $K^0(\mathbb{P}^n)$ can be written as $\mathbb{Z}[\alpha]/(\alpha^{n+1})$, where $\alpha = [\mathcal{O}_H] = 1 - [\mathcal{O}(-1)]$ is the class of the structure sheaf of a hyperplane. As a special case, the K-theory of a point is $\mathbb{Z}$.

If X is a smooth variety equipped with an action of a torus T, we can define its *equivariant K-theory* $K_T^0(X) \cong K_0^T(X)$. The construction and properties are exactly the same as in the previous paragraphs, if we replace "vector bundles" and "coherent sheaves" by "T-equivariant vector bundles" and "T-equivariant coherent sheaves".

For later reference, we describe the equivariant K-theory of a point: $K_T^0(pt) = \mathbb{Z}[\mathrm{Char}(T)]$, where $\mathrm{Char}(T) = \mathrm{Hom}(T, \mathbb{C}^*)$ is the lattice of characters of T. Here $\mathbb{Z}[\mathrm{Char}(T)]$ is the *group ring* of $\mathrm{Char}(T)$, i.e. as a module over $\mathbb{Z}$ it has a basis given by $\mathrm{Char}(T)$, and multiplication is induced from addition in $\mathrm{Char}(T)$. It is isomorphic to the ring of Laurent polynomials in $\dim T$ variables. Explicitly, a T-equivariant sheaf on pt is just a vector space W with a T-action. We may decompose $W = \oplus_{\mathbf{c} \in \mathrm{Char}(T)} W_{\mathbf{c}}$ (see Section 3.2.1). The corresponding element of $\mathbb{Z}[\mathrm{Char}(T)]$ is the character (also called *Hilbert series*) $\mathrm{Hilb}(W) := \sum_{\mathbf{c} \in \mathrm{Char}(T)} (\dim W_{\mathbf{c}}) \mathbf{c}$. We point out that even for infinite-dimensional T-modules, $\mathrm{Hilb}(W)$ makes sense as a formal power series, as long as $W_{\mathbf{c}}$ is finite-dimensional for all $\mathbf{c}$. We finish this section by describing the relation between ordinary and T-equivariant K-theory.

Theorem 6.1.3 ([Mer97, Theorem 4.3]). *Let X be a smooth projective variety with an action of a torus T. Let $S \subseteq T$ be a subtorus. Then the natural map*

$$K_T^0(X) \otimes_{\mathbb{Z}[\mathrm{Char}(T)]} \mathbb{Z}[\mathrm{Char}(S)] \to K_S^0(X)$$

is an isomorphism. In particular, taking S to be the trivial group, the natural map

$$K_T^0(X) \otimes_{\mathbb{Z}[\mathrm{Char}(T)]} \mathbb{Z} \to K^0(X)$$

is an isomorphism.

We note that the map $\mathbb{Z}[\mathrm{Char}(T)] \to \mathbb{Z}$ above is given in coordinates by sending each generator t_i of T to 1.

6.1.2 Explicit construction via equivariant localization

Let X be a smooth projective variety over $\mathbb{C}$, and T a torus acting on it. If X has only finitely many torus-fixed points, we can use the method of *equivariant localization* to give an explicit combinatorial description of classes in $K_T^0(X)$. Our exposition here is largely based on the one in [FS12]. The following theorem is central to our discussion.

Theorem 6.1.4 ([Nie74, Theorem 3.2], see also [FS10, Theorem 2.5] and the references therein). *If X is a smooth projective variety with a torus action, then the restriction map $K_T^0(X) \to K_T^0(X^T)$ is an injection.*

From now on we will always assume that X has only finitely many torus-fixed points. In this case $K_T^0(X^T)$ is simply the ring of functions from X^T to $\mathbb{Z}[\mathrm{Char}(T)]$. In other words, we can describe a class in $K_T^0(X)$ just by giving a finite collection of Laurent polynomials in $\mathbb{Z}[\mathrm{Char}(T)]$.

Remark 6.1.5. In the literature, a variety X whose equivariant K-theory $K_T^0(X)$ is a free $\mathbb{Z}[\mathrm{Char}(T)]$-module, and has a $\mathbb{Z}[\mathrm{Char}(T)]$-basis that restricts to a $\mathbb{Z}$-basis of $K^0(X)$, is called *equivariantly formal*. This notion was first introduced in [GKM98]. In [And12, Section 2.4], it is noted that smooth projective varieties with finitely many T-fixed points are equivariantly formal.

We now explicitly describe the class of a T-equivariant coherent sheaf on X. We will do this under the following additional assumption (which is not essential but makes notation easier and will hold for all varieties of interest).

Definition 6.1.6. A finite-dimensional representation of T is called *contracting* if all characters lie in an open halfspace, or equivalently if the characters generate a pointed cone (see Section 6.2.1). The action of T on a variety X is *contracting*, if for every torus-fixed point $x \in X$, there exists an open neighbourhood U_x isomorphic to $\mathbb{A}^N$ such that the action of T on U_x is a contracting representation.

Let E be a T-equivariant coherent sheaf on X. We will construct a map $[E]^T : X^T \to \mathbb{Z}[\mathrm{Char}(T)]$. For every $x \in X^T$, we have an open neighbourhood U_x as in Definition 6.1.6. Let $\chi_1, \ldots, \chi_N$ be the characters by which T acts on U_x (so $\mathcal{O}(U_x)$ is a polynomial ring multigraded by T in the sense of [MS05, Definition 8.1], with characters $\chi_1^{-1}, \ldots, \chi_N^{-1}$). Our sheaf E, restricted to U_x, corresponds to a graded, finitely generated $\mathcal{O}(U_x)$-module $E(U_x)$.

Since $E(U_x)$ is a graded module over the polynomial ring $\mathcal{O}(U_x)$, which is multigraded by T, it follows from [MS05, Theorem 8.20] that $E(U_x)$ is a T-module, and its Hilbert series is of the form

$$\frac{K(E(U_x), \mathbf{t})}{\prod_{i=1}^{N} (1 - \chi_i^{-1})}, \tag{6.1.1}$$

for some $K(E(U_x), \mathbf{t}) \in \mathbb{Z}[\mathrm{Char}(T)]$.

Definition 6.1.7. For E a T-equivariant coherent sheaf on X, we define $[E]^T : X^T \to \mathbb{Z}[\mathrm{Char}(T)]$ to be the map that sends $x \in X^T$ to $K(E(U_x), \mathbf{t}) \in \mathbb{Z}[\mathrm{Char}(T)]$, the numerator in (6.1.1).

Theorem 6.1.8 (See [FS10, Theorem 2.6], [MS05, Theorem 8.34]). *The map $[E]^T$ defined above is the image of the class of E under the injection $K_T^0(X) \hookrightarrow K_T^0(X^T)$ of Theorem 6.1.4.*

Hence, we can identify $[E]^T : X^T \to \mathbb{Z}[\mathrm{Char}(T)]$ with the T-equivariant K-class of E, and will often write $[E]^T \in K_T^0(X)$.

Example 6.1.9. Let $X = \mathbb{P}^n$, equipped with the natural torus action $\mathbf{t} \cdot [a_0 : \ldots : a_n] = [t_0^{-1} a_0 : \ldots : t_n^{-1} a_n]$. Then $\mathcal{O}(d)$ is a T-equivariant sheaf. The torus action on $\mathbb{P}^n$ has $n + 1$ fixed points, namely $p_i = [0 : \ldots : 1 : \ldots : 0]$, where the 1 is at position i. We use equivariant localization to describe the class $[\mathcal{O}(d)]^T$. Every p_i has an open neighbourhood $U_i = \mathrm{Spec}\, A_i$, where $A_i = \mathbb{C}[x_0, \ldots, \hat{x}_i, \ldots, x_n]$ is multigraded by T via $\deg(x_j) = t_i^{-1} t_j$. The A_i-module $\mathcal{O}(d)(U_i)$ is a copy of A_i generated in degree t_i^d. So its Hilbert series is $t_i^d / \prod_j (1 - t_i^{-1} t_j)$. Hence $[\mathcal{O}(d)]^T$ can be represented by the map $(\mathbb{P}^n)^T \to \mathbb{Z}[\mathrm{Char}(T)] : p_i \mapsto t_i^d$.

We can describe the image of the map from Theorem 6.1.4 explicitly, if we impose an additional condition on X.

Theorem 6.1.10 ([VV03, Corollary 5.12], see also [FS10, Theorem 2.9] and the references therein). *Suppose X is a projective variety with an action of a torus T, such that X has finitely many T-fixed points and finitely many 1-dimensional T-orbits, each of which has closure isomorphic to $\mathbb{P}^1$. Then a map $f : X^T \to \mathbb{Z}[\mathrm{Char}(T)]$ is in the image of the map $K_T^0(X) \to K_T^0(X^T)$ of Theorem 6.1.4 if and only if the following condition holds:*
For every one-dimensional orbit, on which T acts by character χ and for which x and y are the T-fixed points in the orbit closure, we have

$$f(x) \equiv f(y) \mod 1 - \chi. \tag{6.1.2}$$

Remark 6.1.11. Given a variety X as in Theorem 6.1.10, one defines the *moment graph* of X to be the graph whose vertices are the T-fixed points of X, with edges corresponding to the one-dimensional T-orbits and every edge labeled by the relevant character. By the discussion above, the ring $K_T^0(X)$ is determined by the moment graph of X: an element of $K_T^0(X)$ is given by putting a Laurent polynomial $f(x) \in \mathbb{Z}[\mathrm{Char}(T)]$ at every vertex x, such that for every edge the congruence (6.1.2) holds.

Example 6.1.12. We continue Example 6.1.9. Note that $\mathbb{P}^n$ has only finitely many one-dimensional torus orbits: for every pair p_i, p_j of T-fixed points, there is a unique T-orbit whose closure contains p_i and p_j. Furthermore, T acts on this orbit with character $t_j^{-1} t_i$. We see that $t_i^d \equiv t_j^d \mod 1 - t_j^{-1} t_i$, so that the class $[\mathcal{O}(d)]^T$ indeed fulfills the condition of Theorem 6.1.10.

We also can describe pullback and pushforward in the language of equivariant localization. Let $\pi : X \to Y$ be a T-equivariant map of smooth projective varieties with finitely many T-fixed points, then for $[E]^T \in K_T^0(Y)$, its pullback can be computed by

$$(\pi^*[E]^T)(x) = [E]^T(\pi(x)) \tag{6.1.3}$$

for $x \in X^T$.

Describing the pushforward of $[F]^T \in K_T^0(X)$ is a bit more complicated. Suppose that X and Y are contracting. For every point $x \in X^T$ (resp. $y \in Y^T$), we pick as before an open neighbourhood U_x (resp. V_y) on which T acts by characters $\chi_1(x), \ldots, \chi_r(x)$ (resp. $\eta_1(y), \ldots, \eta_s(y)$). Then the pushforward of $[F]^T$ is determined by the formula

$$\frac{(\pi_*[F]^T)(y)}{\prod (1 - \eta_j(y)^{-1})} = \sum_{x \in \pi^{-1}(y) \cap X^T} \frac{[F]^T(x)}{\prod (1 - \chi_i(x)^{-1})}. \tag{6.1.4}$$

For a proof, we refer the reader to [CG10, Theorem 5.11.7].

In the case where Y is a point, the pushfoward $\pi_*[F]^T$ (also called *Lefschetz trace* or *equivariant Euler characteristic*) will be denoted by $\chi([F]^T) \in \mathbb{Z}[\mathrm{Char}(T)]$. Formula (6.1.4) reduces to:

$$\chi([F]^T) = \sum_{x \in X^T} \frac{[F]^T(x)}{\prod (1 - \chi_i(x)^{-1})}. \tag{6.1.5}$$

Remark 6.1.13. We can use Theorem 6.1.3 to obtain a description of the ordinary K-theory using equivariant localization. However, one should be careful when using this for computations in practice. Here is a toy example: let $X = \mathbb{P}^2$ with the usual action of $(\mathbb{C}^*)^2$. Then $X^T = \{[1 : 0], [0 : 1]\}$, and we can write the elements of $K_T^0(X^T) \cong \mathrm{Maps}(X^T, \mathbb{Z}[t_0^{\pm}, t_1^{\pm}]) \ni f$ as pairs $(f([1 : 0]), f([0 : 1]))$. Then $(t_0 - t_1, 0)$ satisfies the condition from Theorem

6.1.10, hence it gives a class in $K_T^0(X)$. It is tempting to do the following computation in $K^0(X) \cong K_T^0(X) \otimes_{\mathbb{Z}[\mathrm{Char}(T)]} \mathbb{Z}$:

$$(t_0 - t_1, 0) \otimes 1 = (1, 0) \otimes (1 - 1) = 0$$

but this is wrong! Indeed, $(1,0)$ does not satisfy the condition from Theorem 6.1.10, hence is not in $K_T^0(X)$. In fact, one can check that $(t_0 - t_1, 0)$ is the equivariant class of the torus-fixed point $[1 : 0] \in \mathbb{P}^2$.

6.2 Equivariant K-theory of Grassmannians and flag varieties

6.2.1 A short review on cones and their Hilbert series

In this subsection, we introduce some notation that will be needed later to define the K-class of a (flag) matroid. For more details, we refer to [CLS11, Section 1.2] and [Sta12, Section 4.5].

Recall that a *convex polyhedral rational cone* is a subset of $\mathbb{R}^n$ of the form

$$C = \{\mathbf{v} + \sum_{i=0}^{m} a_i \mathbf{u}_i \mid a_i \geq 0 \text{ for all } i \in [m]\},$$

where $\mathbf{v}, \mathbf{u}_0, \ldots, \mathbf{u}_{m-1} \in \mathbb{Z}^n \subset \mathbb{R}^n$. A cone is called *pointed* if it does not contain a line. If C is a pointed rational cone with vertex at the origin, then every one-dimensional face ρ contains a unique lattice point $\mathbf{u}_\rho$ that is closest to the origin; we call $\mathbf{u}_\rho$ the *ray generator* of ρ. It is not hard to see that $MG(C) := \{\mathbf{u}_\rho \mid \rho \text{ a one-dimensional face of } C\}$ is a minimal generating set of C. If the vertex $\mathbf{v}$ of C is not at the origin, we define $MG(C) := MG(C - \mathbf{v})$. If the minimal generators are linearly independent over $\mathbb{R}$, we call C *simplicial*. If they are part of a $\mathbb{Z}$-basis of $\mathbb{Z}^n$, we call C *smooth*.

For a pointed cone C in $\mathbb{R}^n$, we define its *Hilbert series* $\mathrm{Hilb}(C)$ by:

$$\mathrm{Hilb}(C) := \sum_{\mathbf{a} \in C \cap \mathbb{Z}^n} \mathbf{t}^{\mathbf{a}}.$$

This is always a rational function, with denominator equal to $\prod_{\mathbf{u} \in MG(C)} (1 - \mathbf{t}^{\mathbf{u}})$ [Sta12, Theorem 4.5.11]. If C is a smooth cone, then its Hilbert series is easy to compute: $\mathrm{Hilb}(C) = \prod_{\mathbf{u} \in MG(C)} \frac{\mathbf{t}^{\mathbf{v}}}{1 - \mathbf{t}^{\mathbf{u}}}$. If C is a simplicial cone, we can compute its Hilbert series as follows. First compute the finite set $D_C := \{\mathbf{b} \in C \cap \mathbb{Z}^n : \mathbf{b} = \sum_{\mathbf{u} \in MG(C)} \lambda_{\mathbf{u}} \mathbf{u} \mid 0 \leq \lambda_{\mathbf{u}} < 1\}$. Then

$$\mathrm{Hilb}(C) = \left(\sum_{\mathbf{b} \in D_C} \mathbf{t}^{\mathbf{b}} \right) \prod_{\mathbf{u} \in MG(C)} \frac{\mathbf{t}^{\mathbf{v}}}{1 - \mathbf{t}^{\mathbf{u}}}.$$

For an arbitrary pointed rational polyhedral cone, we can compute its Hilbert series by triangulating it.

6.2.2 Matroids and the K-theory of Grassmannians

In this subsection we compute the class in equivariant K-theory of a torus orbit closure in a Grassmannian. We then note that this class only depends on the underlying matroid, and give a combinatorial algorithm to get the class in K-theory directly from the matroid. This algorithm can then be used as a definition to associate a class in K-theory to an arbitrary (not necessarily representable) matroid. This was first done by Fink and Speyer in [FS12].

Let us first fix the following sign conventions. The torus $T = (\mathbb{C}^*)^n$ acts on $\mathbb{C}^n$ as follows: $\mathbf{t} \cdot (x_1, \ldots, x_n) = (t_1^{-1}x_1, \ldots, t_n^{-1}x_n)$. The action of T on $Gr(r, n)$ is induced from this action. Explicitly, if $p \in Gr(r, n)$ has Plücker coordinates $[P_I]_{I \in \binom{[n]}{k}}$, then $\mathbf{t} \cdot p$ has Plücker coordinates $[(\prod_{i \in I} t_i^{-1})P_I]_{I \in \binom{[n]}{k}}$. We begin by describing the T-equivariant K-theory of the Grassmannian $Gr(r, n)$ using equivariant localization.

The torus-fixed points of $Gr(r, n)$ are easy to describe: for every size r subset $I \subset [n]$, we define the r-plane $L_I = \operatorname{span}(\{e_i \mid i \in I\}) \subset \mathbb{C}^n$, and denote the corresponding point in $Gr(r, n)$ by p_I. In Plücker coordinates, p_I is given by $P_J = 0$ if $J \neq I$. It is easy to see that the $\binom{n}{r}$ points p_I are precisely the torus-fixed points of $Gr(r, n)$.

We can also describe the one-dimensional torus orbits: there is a (unique) one-dimensional torus orbit between p_I and p_J if and only if $|I \cap J| = r - 1$. In this case, we write $I - J = \{i\}$, $J - I = \{j\}$. If we identify the one-dimensional orbit from p_I to p_J with $\mathbb{A}^1 \setminus 0$ in such a way that the origin corresponds to the torus-fixed point p_I (and so p_J corresponds to the point at infinity), then T acts on the orbit with character $t_j^{-1}t_i$.

Let us now check that the action of T is contracting. We fix a torus-fixed point p_I, and consider the open neighbourhood U_I given by $P_I = 1$. Then $U_I \cong \mathbb{A}^{r(n-r)}$. For $p \in U_I$, we will denote its coordinates with $(u_{i,j})_{i \in I, j \notin I}$, where $u_{i,j} = \frac{P_{I-i \cup j}}{P_I}$. Then $\mathbf{t} \cdot p$ has coordinates $(t_j^{-1}t_i u_{i,j})_{i \in I, j \notin I}$. Thus, T acts on this space with characters $t_j^{-1}t_i$, where $i \in I, j \notin I$. Identifying $t_1^{a_1} \cdots t_n^{a_n}$ with $(a_1, \ldots, a_n)$, all these points lie in the open halfspace defined by $\sum_{i \in I} a_i > 0$.

Example 6.2.1. We compute the class of $\mathcal{O}(1)$. The sheaf $\mathcal{O}(1)$ on $Gr(r, n)$ was already mentioned in Section 5.2: it is the pullback of $\mathcal{O}(1)$ on $\mathbb{P}^{\binom{n}{r}-1}$ via the Plücker embedding. We can also describe $\mathcal{O}(1)$ as $\bigwedge^r \mathcal{S}^\vee$, where $\mathcal{S}$ is the tautological bundle on $Gr(r, n)$ which will be introduced in Example 6.2.2 below.

We can apply Theorem 6.1.3 to the result from Example 6.1.9 to replace the torus action on $\mathbb{P}^{\binom{n}{r}-1}$ with a different torus action, induced from the action on the Plücker coordinates. By applying the pullback formula (6.1.3), we find that the class $[\mathcal{O}(1)]^T$ in equivariant K-theory is given by the map

$$[\mathcal{O}(1)]^T : Gr(r, n)^T \to \mathbb{Z}[\operatorname{Char}(T)] : p_I \mapsto t_{i_1} \cdots t_{i_r},$$

where we wrote $I = \{i_1, \ldots, i_r\}$.

109

Example 6.2.2. Two important vector bundles on $Gr(r,n)$ are the *tautological subbundle* $\mathcal{S}$, whose fiber over a point $L \in Gr(r,n)$ is the r-plane L, and the *tautological quotient bundle* $\mathcal{Q}$, which can by definied via the short exact sequence

$$0 \to \mathcal{S} \to \mathbb{C}^n \to \mathcal{Q} \to 0.$$

The class $[\mathcal{S}]^T \in K_0^T(Gr(r,n))$ can be described by the map

$$[\mathcal{S}]^T : Gr(r,n)^T \to \mathbb{Z}[\mathrm{Char}(T)] : p_I \mapsto \sum_{i \in I} t_i^{-1}.$$

More generally, the class $[\bigwedge^p \mathcal{S}]$ is given by

$$[\overset{p}{\bigwedge} \mathcal{S}]^T : Gr(r,n)^T \to \mathbb{Z}[\mathrm{Char}(T)] : p_I \mapsto \sum_{\mathsf{p} \in \binom{I}{p}} t^{-e_\mathsf{p}}. \tag{6.2.1}$$

There are similar formulas for $[\bigwedge^p \mathcal{S}^\vee]^T$, $[\bigwedge^q \mathcal{Q}]^T$, and $[\bigwedge^q \mathcal{Q}^\vee]^T$: just replace $\sum_{\mathsf{p} \in \binom{I}{p}} t^{-e_\mathsf{p}}$ in (6.2.1) respectively by $\sum_{\mathsf{p} \in \binom{I}{p}} t^{e_\mathsf{p}}$, $\sum_{\mathsf{q} \in \binom{[n] \setminus I}{q}} t^{-e_\mathsf{q}}$, $\sum_{\mathsf{q} \in \binom{[n] \setminus I}{q}} t^{e_\mathsf{q}}$.

Let p be a point in $Gr(r,n)$ and $M = M_p$ be the corresponding matroid on $[n]$. Then $\overline{Tp}$ is a closed subvariety of $Gr(r,n)$; in particular, it is given by a coherent sheaf. We want to compute its class in T-equivariant K-theory, which is a map $[\overline{Tp}]^T : Gr(r,n)^T \to \mathbb{Z}[\mathrm{Char}(T)]$. As before, let $p_I \in Gr(r,n)^T$ be the torus-invariant point given by $P_J = 0$ for $J \neq I$, and let U_I be the affine open neighbourhood U_I of p_I defined by $P_I = 1$. If I is not a basis of M, then $\overline{Tp}$ does not intersect U_I, hence $[\overline{Tp}]^T(p_I) = 0$. Hence, we will assume that I is a basis of M, i.e. that $p \in U_I$.

The coordinate ring of $\overline{Tp} \cap U_I$ is isomorphic to $\mathbb{C}[s_i^{-1}s_j]$, where $s_i^{-1}s_j$ is a generator if and only if $(I-i)\cup j$ is a basis of M. We will denote this ring by $R_{M,I}$. This ring should be viewed as a T-module, with $\mathbf{t}\cdot s_i^{-1}s_j = t_i^{-1}t_j s_i^{-1}s_j$. The Hilbert series of $R_{M,I}$ is a rational function with denominator dividing $\prod_{i \in I}\prod_{j \notin I}(1 - t_i^{-1}t_j)$. Thus, by (6.1.1),

$$[\overline{Tp}]^T(p_I) = \mathrm{Hilb}(R_{M,I})\prod_{i \in I}\prod_{j \notin I}(1 - t_i^{-1}t_j). \tag{6.2.2}$$

Definition 6.2.3. For any lattice polytope P and v a vertex of P, we define $\mathrm{Cone}_v(P)$, also known as the tangent cone of P at the vertex v, to be the cone spanned by all vectors of the form $u - v$ with $u \in P$. For $I \in \binom{[n]}{k}$, we write $\mathrm{Cone}_I(M) := \mathrm{Cone}_{e_I}(P(M))$ if I is a basis of M, and $\mathrm{Cone}_I(M) := \emptyset$ otherwise.

Since $\mathrm{Cone}_I(M)$ is the positive real span of all vectors $e_J - e_I$, where $J \in \mathcal{B}(M)$, we find that $\mathrm{Hilb}(R_{M,I}) = \mathrm{Hilb}(\mathrm{Cone}_I(M))$. So (6.2.2) can also be written as

$$[\overline{Tp}]^T(p_I) = \mathrm{Hilb}(\mathrm{Cone}_I(M))\prod_{i \in I}\prod_{j \notin I}(1 - t_i^{-1}t_j). \tag{6.2.3}$$

We note that the right hand side of (6.2.3) only depends on the matroid M and not on the chosen point p or even the torus orbit $\overline{Tp}$. Moreover, the formulas make sense even for non-representable matroids. Thus we can use them as a definition for the class in K-theory for a matroid.

Definition 6.2.4 ([FS12]). For any rank k matroid M on $[n]$, we define $y(M) : Gr(r, n)^T \to \mathbb{Z}[\mathrm{Char}(T)]$ by

$$y(M)(p_I) = \mathrm{Hilb}(\mathrm{Cone}_I(M)) \prod_{i \in I} \prod_{j \notin I} (1 - t_i^{-1} t_j)$$

if I is a basis of M, and $y(M)(p_I) = 0$ otherwise.

Theorem 6.2.5 ([FS12, Proposition 3.2]). *The class $y(M) \in K_T^0(Gr(r, n)^T)$ satisfies the condition of Theorem 6.1.10, and hence defines a class in $K_T^0(Gr(r, n))$.*

6.2.3 Flag matroids and the K-theory of flag varieties

In this subsection, we generalize the results from the previous subsection replacing "matroids" by "flag matroids" and "Grassmannians" by "flag varieties". We first describe the equivariant K-theory of a flag variety $Fl(\mathbf{r}, n)$. The torus-fixed points are given as follows: for every (set-theoretic) flag $F = (F_1 \subseteq \ldots \subseteq F_k)$ of rank $\mathbf{r}$ on $[n]$, we define a (vector space) flag $L_1 \subseteq \ldots \subseteq L_k$ by $L_i = \mathrm{span}(\{e_j \mid j \in F_i\})$. We will denote the corresponding point in $Fl(\mathbf{r}, n)$ by p_F. The Plücker coordinates of p_F are given by $P_S = 1$ if S is a constituent of F and $P_S = 0$ otherwise. Here, the Plücker coordinates of a point in $Fl(\mathbf{r}, n)$ are the ones induced from the embedding $Fl(\mathbf{r}, n) \hookrightarrow \prod Gr(r_i, n)$.

We can also describe the one-dimensional torus orbits: let p_F be a torus-fixed point. We define $\mathrm{Ex}(F)$ to be the set of all pairs $(i, j) \in [n] \times [n]$ for which there exists an ℓ such that $i \in F_\ell$ and $j \notin F_\ell$. For every $(i, j) \in \mathrm{Ex}(F)$, we define a new flag $F' = F_{i \to j}$ by switching the roles of i and j. More precisely: if $i \in F_\ell$ but $j \notin F_\ell$, then $F'_\ell := (F_\ell - i) \cup j$; in any other case $F'_\ell := F_\ell$. Then there is a unique one-dimensional torus orbit between p_F and $p_{F'}$, and all one-dimensional torus orbits arise in this way. The torus T acts on this orbit with character $t_j^{-1} t_i$.

Lemma 6.2.6. *The action of T on $Fl(\mathbf{r}, n)$ is contracting.*

Proof. For every torus-fixed point p_F, we consider the open neighbourhood U_F given by $P_{F_\ell} \neq 0$ for all ℓ. Then $U_F \cong \mathbb{A}^N$, where $N = \dim(Fl(\mathbf{r}, n)) = \sum_{i=1}^k r_i(r_{i+1} - r_i)$ (here $r_{k+1} := n$). We will denote the coordinates of a point q in U_F by $(u_{i,j})_{(i,j) \in \mathrm{Ex}(F)}$, where $u_{i,j} = \frac{P_{F_\ell - i \cup j}}{P_{F_\ell}}$ for any r which satisfies $i \in F_\ell$ and $j \notin F_\ell$. Then $\mathbf{t} \cdot q$ has coordinates $(t_j^{-1} t_i u_{i,j})_{(i,j) \in \mathrm{Ex}(F)}$. So T acts on U_F with characters $t_j^{-1} t_i$, $(i, j) \in \mathrm{Ex}(F)$. As before, identifying $t_1^{a_1} \cdots t_n^{a_n}$ with $(a_1, \ldots, a_n)$, all these characters lie on the open halfspace $\sum_{\ell=1}^k \sum_{i \in F_\ell} a_i > 0$. $\qquad\square$

111

Example 6.2.7. For every $i \in [k]$, we have the following tautological exact sequence of vector bundles on $Fl(\mathbf{r}, n)$:

$$0 \to \mathcal{S}_i \to \mathbb{C}^n \to \mathcal{Q}_i \to 0, \tag{6.2.4}$$

where $\mathcal{S}_i$ is the vector bundle whose fiber at a point $\mathbf{L} = (L_1, \ldots, L_k) \in Fl(\mathbf{r}; n)$ is the subspace L_i. Similar to Example 6.2.2, the class $[\bigwedge^p \mathcal{S}_i]$ is given by

$$[\bigwedge^p \mathcal{S}_i]^T : Fl(\mathbf{r}, n)^T \to \mathbb{Z}[\mathrm{Char}(T)] : p_F \mapsto \sum_{\mathfrak{p} \in \binom{F_i}{p}} t^{-e_{\mathfrak{p}}}. \tag{6.2.5}$$

The analoguous formulas for $[\bigwedge^p \mathcal{S}_i^{\vee}]^T$, $[\bigwedge^q \mathcal{Q}_i]^T$, $[\bigwedge^q \mathcal{Q}_i^{\vee}]^T$ hold as well. For $\mathbf{a} = (a_1, \ldots, a_k) \in \mathbb{Z}^k$ we denote by $\mathcal{O}(\mathbf{a})$ the line bundle $\bigotimes_{i=1}^k (\det \mathcal{S}_i^{\vee})^{\otimes a_i}$, see also Remark 5.4.1. The class $[\mathcal{O}(\mathbf{a})]^T$ is given by sending p_F to the character $t^{a_1 e_{F_1} + \cdots + a_k e_{F_k}}$.

Let p be a point in $Fl(\mathbf{r}, n)$, and let $\mathcal{F}$ be the corresponding flag matroid on $[n]$. We want to compute the T-equivariant class of $\overline{Tp}$, which is a map $[\overline{Tp}]^T : Fl(\mathbf{r}, n)^T \to \mathbb{Z}[\mathrm{Char}(T)]$. We fix a point $p_F \in Fl(\mathbf{r}, n)^T$, and consider the affine neighbourhood U_F described above. If F is not a basis of $\mathcal{F}$, then $\overline{Tp}$ does not intersect U_F, hence $[\overline{Tp}]^T(p_F) = 0$. Thus, we will assume that F is a basis of $\mathcal{F}$, i.e. that $p \in U_F$.

The coordinate ring of $\overline{Tp} \cap U_F$ is isomorphic to $k[s_i^{-1} s_j]$, where $s_i^{-1} s_j$ is a generator if and only if $F_{i \to j} \in \mathcal{F}$. We will denote this ring by $R_{\mathcal{F}, F}$. This ring should be viewed as a T-module, with $\mathbf{t} \cdot s_i^{-1} s_j = t_i^{-1} t_j s_i^{-1} s_j$. The Hilbert series of this T-module is a rational function with denominator dividing $\prod_{(i,j) \in \mathrm{Ex}(F)} (1 - t_i^{-1} t_j)$. Thus, by (6.1.1),

$$[\overline{Tp}]^T(p_F) = \mathrm{Hilb}(R_{\mathcal{F}, F}) \prod_{(i,j) \in \mathrm{Ex}(F)} (1 - t_i^{-1} t_j). \tag{6.2.6}$$

Definition 6.2.8. We write $\mathrm{Cone}_F(\mathcal{F})$ for $\mathrm{Cone}_{e_F}(P(\mathcal{F}))$, as in Definition 6.2.3.

As before, we find that $\mathrm{Hilb}(R_{\mathcal{F}, F}) = \mathrm{Hilb}(\mathrm{Cone}_F(\mathcal{F}))$. Hence, (6.2.6) can also be written as

$$[\overline{Tp}]^T(p_F) = \mathrm{Hilb}(\mathrm{Cone}_F(\mathcal{F})) \prod_{(i,j) \in \mathrm{Ex}(F)} (1 - t_i^{-1} t_j). \tag{6.2.7}$$

Again, (6.2.7) only depends on the flag matroid $\mathcal{F}$ and not on the chosen point p or even the torus orbit $\overline{Tp}$. Moreover the formulas make sense even for non-representable flag matroids. Thus we can use them as a definition for the class in K-theory for a flag matroid.

Definition 6.2.9. For any rank $\mathbf{r}$ flag matroid $\mathcal{F}$ on $[n]$, we define $y(\mathcal{F})$: $Fl(\mathbf{r}, n)^T \to \mathbb{Z}[\mathrm{Char}(T)]$ by

$$y(\mathcal{F})(p_F) = \mathrm{Hilb}(\mathrm{Cone}_F(\mathcal{F})) \prod_{(i,j) \in \mathrm{Ex}(F)} (1 - t_i^{-1} t_j)$$

if F is a basis of $\mathcal{F}$, and $y(\mathcal{F})(p_F) = 0$ otherwise.

Proposition 6.2.10. *The class* $y(\mathcal{F}) \in K_T^0(Fl(\mathbf{r}, n)^T)$ *satisfies the condition of Theorem 6.1.10, and hence defines a class in* $K_T^0(Fl(\mathbf{r}, n))$.

Proof. The proof is a straightforward generalization of the proof of [FS12, Proposition 3.2]. $\qquad\square$

Example 6.2.11. Let $\mathcal{F}$ be the flag matroid of Example 5.3.14. We first compute $y(\mathcal{F})(p_F)$, where F is the flag $2 \subseteq 12$ (so $e_F = (1, 2, 0)$). From Figure 5.2, it is clear that $\mathrm{Cone}_F(\mathcal{F}) = \mathrm{Cone}((1, -1, 0), (0, -1, 1))$, which has Hilbert series $\frac{1}{(1 - t_2^{-1} t_1)(1 - t_2^{-1} t_3)}$. Furthermore, we have $\mathrm{Ex}(F) = \{(2, 1), (2, 3), (1, 3)\}$. We find that $y(\mathcal{F})(p_F) = 1 - t_1^{-1} t_3$. We can do the same for the other torus-fixed points. The result is summarized in Figure 6.1.

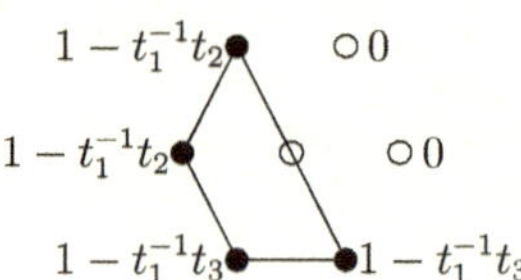

Figure 6.1: The class in K-theory of a flag matroid

6.2.4 The Tutte polynomial via K-theory

In [FS12], a geometric description of the Tutte polynomial of a matroid is given. Consider the diagram below, where all the maps are natural projections or inclusions. The inclusion on the right is the composition $Fl(1, n - 1; n) \hookrightarrow Gr(n - 1, n) \times Gr(1, n) \cong (\mathbb{P}^{n-1})^\vee \times \mathbb{P}^{n-1}$.

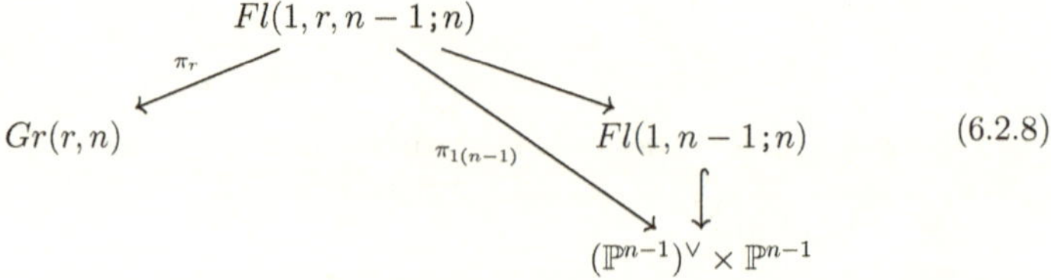

$$(6.2.8)$$

Generalizing Example 6.1.2, one can show that

$$K^0((\mathbb{P}^{n-1})^\vee \times \mathbb{P}^{n-1}) \cong \mathbb{Z}[\alpha, \beta]/(\alpha^n, \beta^n),$$

where $\alpha = [\mathcal{O}_{H_1}]$ is the K-class of the structure sheaf of a hyperplane in $(\mathbb{P}^{n-1})^\vee$ and $\beta = [\mathcal{O}_{H_2}]$ the likewise K-class from $\mathbb{P}^{n-1}$.

Theorem 6.2.12 ([FS12, Theorem 5.1]). *The following equality holds:*

$$(\pi_{1(n-1)})_* \pi_r^*(Y(M) \cdot [\mathcal{O}(1)]) = T_M(\alpha, \beta), \tag{6.2.9}$$

where $Y(M)$ is the class associated to the matroid M in the non-equivariant K-theory of the Grassmannian, and T_M is the Tutte polynomial of M.

In other words, the Tutte polynomial of a matroid can be viewed as a Fourier-Mukai transform of its associated class in K-theory. We can now now generalize this construction to get a *definition* of the Tutte polynomial of a flag matroid.

Definition 6.2.13. Consider the following diagram:

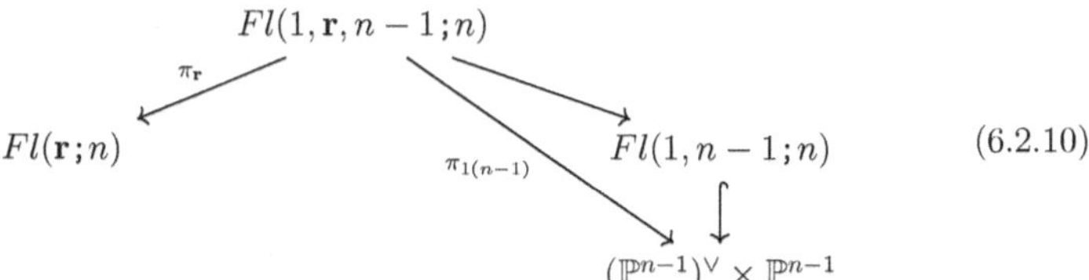

$$\tag{6.2.10}$$

Let $\mathcal{F}$ be a flag matroid on $[n]$ of rank $\mathbf{r}$, and let $Y(\mathcal{F}) \in K^0(Fl(\mathbf{r}; n))$ be its class in non-equivariant K-theory, as in Definition 6.2.9. The *flag-geometric Tutte polynomial* of $\mathcal{F}$, denoted $K\mathcal{T}_{\mathcal{F}}(x, y)$, is the unique bivariate polynomial in x, y of bi-degree at most $(n-1, n-1)$ such that

$$K\mathcal{T}_{\mathcal{F}}(\alpha, \beta) := (\pi_{1(n-1)})_* \pi_{\mathbf{r}}^*(Y(\mathcal{F}) \cdot [\mathcal{O}(1)]). \tag{6.2.11}$$

We computed the flag-geometric Tutte polynomial for some small examples using Sage [S+17], Macaulay2 [GS], and Normaliz [BIR+]. Our code is available at [Sey]. The program first computes the equivariant class $(\pi_{1(n-1)})_* \pi_d^*(y(M) \cdot [\mathcal{O}(1)]) \in K_T^0((\mathbb{P}^{n-1})^\vee \times \mathbb{P}^{n-1})$ using equivariant localization, and then computes the underlying non-equivariant class. In the next chapter, we will introduce faster ways of computing the K-Tutte polynomial.

Example 6.2.14. We consider again the flag matroid of Examples 5.3.14 and 6.2.11. We first compute $y(\mathcal{F}) \cdot [\mathcal{O}(1)]$, which is displayed in figure 6.2. The two projections from $Fl(1, \mathbf{r}, 2; 3) = Fl(1, 1, 2, 2; 3)$ to $Fl(1, 2; 3)$ are isomorphisms, hence pulling back and pushing forward along them does nothing. Next we

114

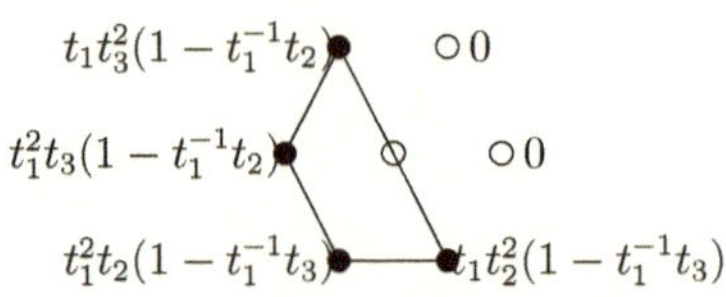

Figure 6.2: $y(\mathcal{F}) \cdot [\mathcal{O}(1)]$

need to push our class $X = y(\mathcal{F}) \cdot [\mathcal{O}(1)] \in K_T^0(Fl(1,2;3))$ to a class $Z \in K_T^0(\mathbb{P}^{n-1} \times \mathbb{P}^{n-1})$, using formula (6.1.4).

The T-fixed points of $\mathbb{P}^{n-1} \times \mathbb{P}^{n-1}$ are given by pairs $p = (\ell, H)$, where $\ell \in Gr(1,3)^T = \{\langle e_1\rangle, \langle e_2\rangle, \langle e_3\rangle\}$ and $H \in Gr(2,3)^T = \{\langle e_1, e_2\rangle, \langle e_1, e_3\rangle, \langle e_2, e_3\rangle\}$. If $\ell \not\subset H$, then $Z(p) = 0$. If $\ell \subset H$, then $p \in Fl(1,2;3) \subset \mathbb{P}^2 \times \mathbb{P}^2$. Since we are pushing forward along an embedding, the formula (6.1.4) has a simple form: we can find characters $\chi_1, \chi_2, \chi_3, \eta$ and open neighbourhoods $p \ni U_1 \subset Fl(1,2;3)$ and $p \ni U_2 \subset \mathbb{P}^2 \times \mathbb{P}^2$, such that T acts on U_1 with characters χ_1, χ_2, χ_3, and on U_2 with characters $\chi_1, \chi_2, \chi_3, \eta$. Then (6.1.4) becomes:

$$Z(p) = (1 - \eta^{-1})X(p).$$

Consider for example $p = (\langle e_1\rangle, \langle e_1, e_2\rangle)$. Then $p \in Fl(1,2;3)$ has an open neighbourhood where T acts by characters $t_2 t_3^{-1}, t_1 t_3^{-1}, t_1 t_2^{-1}$, while $p \in \mathbb{P}^2 \times \mathbb{P}^2$ has an open neighbourhood where T acts by characters $t_2 t_3^{-1}, t_1 t_3^{-1}, t_1 t_3^{-1}, t_1 t_2^{-1}$. We compute that

$$Z((\langle e_1\rangle, \langle e_1, e_2\rangle)) = t_1^2 t_2 (1 - t_1^{-1}t_3)(1 - t_1^{-1}t_3) = t_2(t_1 - t_3)^2.$$

Similarily, we find that

$$Z((\langle e_1\rangle, \langle e_1, e_3\rangle)) = t_3(t_1 - t_2)^2,$$
$$Z((\langle e_3\rangle, \langle e_1, e_3\rangle)) = t_3(t_1 - t_2)(t_3 - t_2),$$
$$Z((\langle e_2\rangle, \langle e_1, e_2\rangle)) = t_2(t_1 - t_3)(t_2 - t_3),$$

and $Z(p) = 0$ in all other cases.

Finally, we need to find the underlying class in non-equivariant K-theory. This is quite tedious to do by hand, so we just refer to the algorithm provided in the Sage code available at [Sey] for this. In the end, we find that

$$K\mathcal{T}_{\mathcal{F}}(x, y) = x^2 y^2 + x^2 y + x y^2 + x^2 + x y.$$

We will revisit this example in the next chapter.

Example 6.2.15. As another example, consider the uniform flag matroid $\mathcal{U}_{(2,3);5}$ of rank $(2, 3)$ on $[5]$ (that is, the constituents of $\mathcal{U}_{(2,3);5}$ are the uniform matroids

$U_{2,5}$ and $U_{3,5}$). Using our program, we find that its flag-geometric Tutte polynomial $K\mathcal{T}_{\mathcal{U}_{(2,3):5}}(x,y)$ equals

$$x^3y^3+2x^3y^2+2x^2y^3+3x^3y+8x^2y^2+3xy^3+4x^3+8x^2y+8xy^2+4y^3+2x^2+4xy+2y^2.$$

Remark 6.2.16. Unlike the usual Tutte polynomial for matroids, the flag-geometric Tutte polynomial need not have nonnegative coefficients. For instance, the flag-geometric Tutte polynomial of the uniform flag matroid $\mathcal{U}_{(1,3):5}$ equals

$$x^3y^4 + x^3y^3 + 2x^2y^4 + x^3y^2 - x^2y^3 + 3xy^4 + x^3y + 6x^2y^2 + 9xy^3 + 4y^4 + x^3 + 3x^2y + 3xy^2 + y^3.$$

The flag-geometric Tutte polynomial is a new invariant of flag matroids, whose definition was motivated by geometry. However, so far we haven't stated any combinatorial properties of this invariant, and we have no efficient way of computing it yet. This will be remedied in the next chapter.

Chapter 7

The flag-geometric Tutte polynomial of a flag matroid

In the end of the previous chapter, we generalized the construction of Fink and Speyer to define the flag-geometric Tutte polynomial $KT_{\mathcal{F}}$ of a flag matroid. In this chapter we will study some combinatorial properties of this invariant.

There is no known corank-nullity type formula for $KT_{\mathcal{F}}$. However, it is possible to prove a formula (Theorem 7.1.4) similar to [FS12, (7)], which will give us an explicit way to compute the flag-geometric Tutte polynomial by summing up Hilbert series of cones. In Section 7.1 we will state and prove the formula; and apply it to prove some first properties of $KT_{\mathcal{F}}$. In Section 7.2, we collect some useful results concerning summations of rational functions arising from lattice points on polyhedra. Our main contribution here is Theorem 7.2.8, which serves as a key technical tool in this chapter and may be of independent interest in the study of lattice polyhedra. In Section 7.3 we apply the methods of Section 7.2 to our formula (7.1.7) in order to prove several combinatorial statements about the K-theoretic Tutte polynomial. We restrict ourselves to the case of 2-step flag matroids; i.e. $\mathcal{F} = (M_1, M_2)$. Our main results are a formula for $KT_{\mathcal{F}}(2,2)$, and a deletion-contraction-like relation for the flag-geometric Tutte polynomial of a flag matroid of the form (M, M). The flag-geometric Tutte polynomial remains a quite mysterious object. In Section 7.4, we list several open problems, and mention some related invariants of flag matroids which could be interesting to study in future work. This chapter is based on the paper [DES20], which is joint work with Rodica Dinu and Christopher Eur.

Computation

At `https://github.com/chrisweur/kTutte`, the reader can find a Macaulay2 [GS] code for computations with torus-equivariant K-classes and flag matroids. In particular, it computes the polynomial $KT_{\mathcal{F}}$ and its torus-equivariant version.

7.1 A formula for the flag-geometric Tutte polynomial

The main result of this section is Theorem 7.1.4. It gives a formula for $K\mathcal{T}_{\mathcal{F}}$ in terms of Hilbert functions of cones. This formula is both useful for proving results about the flag-geometric Tutte polynomial, and for computing it in practice.

7.1.1 The T-equivariant Tutte polynomial

Recall the diagram (6.2.10):

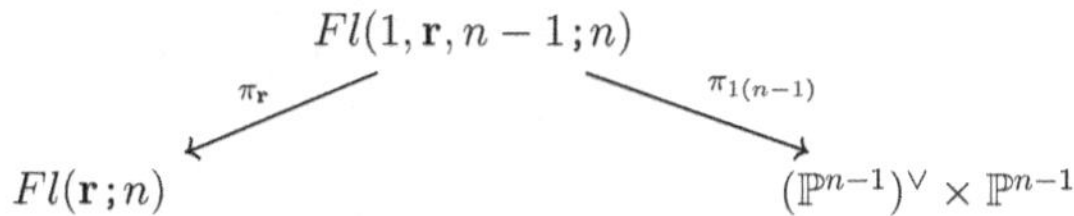

As before, we write $K^0((\mathbb{P}^{n-1})^\vee \times \mathbb{P}^{n-1}) \cong \mathbb{Z}[\alpha, \beta]/(\alpha^n, \beta^n)$, where α and β are the structure sheaves of hyperplanes.

Proposition 7.1.1. *Let $\epsilon \in K_0(Fl(\mathbf{r}; n))$. With u, v as formal variables, define polynomials*

$$R_\epsilon(u, v) := \sum_{p,q} \chi\Big(\epsilon \cdot [\bigwedge^p \mathcal{S}_k][\bigwedge^q \mathcal{Q}_1^\vee]\Big) u^p v^q,$$

where χ denotes the pushforward to a point. Then we have the following identities in $K_0((\mathbb{P}^{n-1})^\vee \times \mathbb{P}^{n-1})$.

$$R_\epsilon(\alpha - 1, \beta - 1) = (\pi_{(n-1)1})_* \pi_{\mathbf{r}}^*(\epsilon).$$

When $k = 1$ (i.e. $Fl(\mathbf{r}; n)$ is a Grassmannian), Proposition 7.1.1 reduces to [FS12, Lemma 4.1]. We remark that, just as in [FS12], Proposition 7.1.1 is an identity in the *non*-equivariant K-theory.

The proof of Proposition 7.1.1 is essentially identical to the proof of [FS12, Lemma 4.1]. Here, as a lemma, we separate out (and also fix a minor error in) the part of the proof in [FS12] that needs modification.

Lemma 7.1.2. *Denote $\eta_1 := (1 - \alpha)^{-1} = [\mathcal{O}(H_1)]$ and $\eta_2 := (1 - \beta)^{-1} = [\mathcal{O}(H_2)]$, and let t be a formal variable. Then the following identities hold in $K_0(Fl(\mathbf{r}; n))[[t]]$:*

$$\sum_p [\bigwedge^p \mathcal{S}_k] t^p = (1 + t)^n (\pi_{\mathbf{r}})_* \pi_{(n-1)1}^* \Big(\frac{1}{1 + t\eta_1}\Big) \tag{7.1.1}$$

$$\sum_q [\bigwedge^q \mathcal{Q}_1^\vee] t^q = (1 + t)^n (\pi_{\mathbf{r}})_* \pi_{(n-1)1}^* \Big(\frac{1}{1 + t\eta_2}\Big). \tag{7.1.2}$$

118

Proof. For each $i = 1, \ldots, k$ note that

$$\left(\sum_{\ell}[\bigwedge^{\ell} \mathcal{S}_i]t^{\ell}\right)\left(\sum_{m}[\bigwedge^{m} \mathcal{Q}_i]t^m\right) = (1+t)^n, \qquad (7.1.3)$$

which follows from the short exact sequence (6.2.4) and [Eis95, A2.2.(c)]. We also have an identity

$$\left(\sum_{\ell}[\bigwedge^{\ell} \mathcal{S}_i]t^{\ell}\right)\left(\sum_{m}(-1)^m[\mathrm{Sym}^m \mathcal{S}_i]t^m\right) = 1 \qquad (7.1.4)$$

and likewise identities for $\mathcal{Q}_i$ and the duals $\mathcal{S}_i^{\vee}, \mathcal{Q}_i^{\vee}$, which follow from the exactness of the Koszul complex [Eis95, A2.6.1]. Now, we note by [Har77, Exercise III.8.4] that

$$(\pi_{\mathbf{r}})_*\pi^*_{(n-1)1}(\eta_2^{\ell}\eta_1^m) = [\mathrm{Sym}^{\ell} \mathcal{S}_1^{\vee} \otimes \mathrm{Sym}^m \mathcal{Q}_k] \qquad (7.1.5)$$

Combining (7.1.3), (7.1.4), and (7.1.5) then yields the desired identities. $\qquad \square$

Sketch of proof of Proposition 7.1.1. Combine Lemma 7.1.2 with the projection formula for K-theory [Ful98, §15.1]. Then expand the power series in u, v, which is in fact a finite sum, and then compare coefficients. See the proof in [FS12, Lemma 4.1] for details. $\qquad \square$

By Proposition 7.1.1, we have that

$$KT_{\mathcal{F}}(u+1, v+1) = \sum_{p,q} \chi\left(y(\mathcal{F})[\mathcal{O}(1)][\bigwedge^{p} \mathcal{S}_k][\bigwedge^{q} \mathcal{Q}_1^{\vee}]\right)u^p v^q,$$

which leads us to the following T-equivariant version of $KT_{\mathcal{F}}$.

Definition 7.1.3. The T-equivariant Tutte polynomial of a flag matroid $\mathcal{F}$ is

$$KT_{\mathcal{F}}^T(u+1, v+1) := \sum_{p,q} \chi^T\left(y(\mathcal{F})^T[\mathcal{O}(1)]^T[\bigwedge^{p} \mathcal{S}_k]^T[\bigwedge^{q} \mathcal{Q}_1^{\vee}]^T\right)u^p v^q. \qquad (7.1.6)$$

We can use (6.1.5) to write $KT_{\mathcal{F}}^T$ as a sum of rational functions as follows. We will use the notation $\mathrm{Hilb}_F(\mathcal{F})$ as shorthand for $\mathrm{Hilb}(\mathrm{Cone}_F(\mathcal{F}))$.

Theorem 7.1.4. *Let $\mathcal{F} = (M_1, \ldots, M_k)$ be a flag matroid on a ground set $[n]$ of rank $\mathbf{r} = (r_1, \ldots, r_k)$. We have*

$$KT_{\mathcal{F}}^T(u+1, v+1) = \sum_{\substack{F = (B_1, \ldots, B_k), \\ F \in \mathcal{B}(\mathcal{F})}} \mathrm{Hilb}_F(\mathcal{F}) \sum_{\mathfrak{p} \subseteq B_k} \sum_{\mathfrak{q} \subseteq [n]\backslash B_1} t^{e_{B_1} + \cdots + e_{B_{k-1}} + e_{\mathfrak{p}} + e_{\mathfrak{q}}} u^{r_k - |\mathfrak{p}|} v^{|\mathfrak{q}|}.$$

$$(7.1.7)$$

Proof. By the pushforward formula (6.1.5), $K\mathcal{T}_{\mathcal{F}}^{T}(u+1, v+1)$ can be written as

$$\sum_{p,q} \sum_{p_F \in Fl(\mathbf{r},n)^T} \frac{y(\mathcal{F})^T(p_F) \cdot [\mathcal{O}(1)]^T(p_F) \cdot [\bigwedge^p \mathcal{S}_k]^T(p_F) \cdot [\bigwedge^q \mathcal{Q}_1^{\vee}]^T(p_F)}{\prod_{(i,j) \in \mathrm{Ex}(F)} (1 - t_i^{-1} t_j)} u^p v^q.$$

The result now follows from Definition 6.2.9 and Example 6.2.7. $\qquad\square$

Many of our results on $K\mathcal{T}_{\mathcal{F}}$ will be obtained by manipulation with the equation (7.1.7).

Example 7.1.5. We once more revisit the flag matroid from Examples 5.3.14, 6.2.11, and 6.2.14. The term in (7.1.7) corresponding to $e_F = (1,2,0)$ is equal to:

$$\frac{t_2(u^2 + t_1 u + t_2 u + t_1 t_2)(1 + t_1 v + t_3 v + t_1 t_3 v^2)}{(1 - t_2^{-1} t_1)(1 - t_2^{-1} t_3)}.$$

Summing up this term with the three terms corresponding to the other bases and expanding the rational function, we find that $K\mathcal{T}_{\mathcal{F}}^{T}(u+1, v+1)$ is given by the following expression:

$$t_1 t_2 t_3 u^2 v^2 + (t_1 t_2 + t_1 t_3 + t_2 t_3) u^2 v + (t_1^2 t_2 t_3 + t_1 t_2^2 t_3 + t_1 t_2 t_3^2) u v^2$$
$$+ (t_1 + t_2 + t_3) u^2 + (t_1^2 t_2 + t_1 t_2^2 + t_1^2 t_3 + 3 t_1 t_2 t_3 + t_2^2 t_3 + t_1 t_3^2 + t_2 t_3^2) u v$$
$$+ (t_1^2 t_2^2 t_3 + t_1^2 t_2 t_3^2) v^2 + (t_1^2 + 2 t_1 t_2 + t_2^2 + 2 t_1 t_3 + t_2 t_3 + t_3^2) u$$
$$+ (t_1^2 t_2^2 + 2 t_1^2 t_2 t_3 + t_1 t_2^2 t_3 + t_1^2 t_3^2 + t_1 t_2 t_3^2) v + (t_1^2 t_2 + t_1 t_2^2 + t_1^2 t_3 + t_1 t_2 t_3 + t_1 t_3^2).$$

Substituting $t_1 = t_2 = t_3 = 1$, $u = x - 1$ and $v = y - 1$, we obtain the non-equivariant flag-geometric Tutte polynomial from Example 6.2.14.

Our Macaulay2 code contains an algorithm that computes the (T-equivariant) flag-geometric Tutte polynomial of a given flag matroid using Theorem 7.1.4, similar to the example above.

7.1.2 First properties

Recall that $U_{0,n}$ is the matroid on $[n]$ whose only basis is the empty set. For any matroid M on $[n]$, the pair $(U_{0,n}, M)$ defines a flag matroid. As a first example, we compute the flag-geometric Tutte polynomial $K\mathcal{T}_{(U_{0,n},M)}$. The diagram (6.2.10) makes sense only when $r_1 \geq 1$, so $K\mathcal{T}_{(U_{0,n},M)}$ cannot be defined as a push-pull of a K-class. Instead, we define $K\mathcal{T}_{(U_{0,n},M)}$ by specializing $K\mathcal{T}_{(U_{0,n},M)}^{T}$ at $t_i = 1$.

Example 7.1.6. For any matroid M on $[n]$, it holds that $KT_{(U_{0,n},M)}(x,y) = y^n T_M(x,1)$. To verify this, we compute

$$KT^T_{(U_{0,n},M)}(u+1,v+1) = \sum_{B\in\mathcal{B}(M)} \mathrm{Hilb}_B(M) \sum_{\mathfrak{p}\subseteq B} \sum_{\mathfrak{q}\subseteq[n]} \mathbf{t}^{e_\mathfrak{p}+e_\mathfrak{q}} u^{r-|\mathfrak{p}|} v^{|\mathfrak{q}|}$$

$$= \Big(\prod_{i=1}^n (1+t_i v)\Big) \cdot \sum_{B\in\mathcal{B}(M)} \mathrm{Hilb}_B(M) \sum_{\mathfrak{p}\subseteq B} \mathbf{t}^{e_\mathfrak{p}} u^{r-|\mathfrak{p}|}$$

$$= \Big(\prod_{i=1}^n (1+t_i v)\Big) \cdot KT^T_M(u+1,1).$$

$$(7.1.8)$$

Setting $t_i = 1$, $u = x - 1$, and $v = y - 1$ yields the desired claim. This example shows that we cannot recover T_M from $KT_{(U_{0,n},M)}$ although $U_{0,n} \twoheadleftarrow M$ is a canonical matroid quotient of M.

We now state some first properties of $KT^T_{\mathcal{F}}$, analogous to Proposition 5.1.14 for the usual Tutte polynomial. Note that if $\mathcal{F}' = (M'_1,\ldots,M'_k)$ and $\mathcal{F}'' = (M''_1,\ldots,M''_k)$ are flag matroids with the same number of constituents on disjoint ground sets, we can define the direct sum $\mathcal{F}' \oplus \mathcal{F}'' := (M'_1 \oplus M''_1,\ldots,M'_k \oplus M''_k)$.

Proposition 7.1.7. *Let $\mathcal{F}$ be a flag matroid on $[n]$. The following properties hold for $KT^T_{\mathcal{F}}$:*

1. *(Direct sum) If $\mathcal{F}$ is a direct sum $\mathcal{F}' \oplus \mathcal{F}''$ of two flag matroids on ground sets A, B with $A \sqcup B = [n]$, then $KT^T_{\mathcal{F}}(x,y) = KT^{T'}_{\mathcal{F}'}(x,y) \cdot KT^{T''}_{\mathcal{F}''}(x,y)$ (where $T' = (\mathbb{C}^*)^A, T'' = (\mathbb{C}^*)^B$).*

2. *(Loops & coloops) Let ℓ be the number of loops in M_1, and c the number of coloops in M_k. Then $x^c y^\ell$ divides $KT_{\mathcal{F}}(x,y)$.*

3. *(Duality) If $\mathcal{F}^\vee$ is the dual flag matroid of $\mathcal{F}$, whose constituents are matroid duals of the original, then $KT_{\mathcal{F}}(y,x) = KT_{\mathcal{F}^\vee}(x,y)$.*

Proof. The first two statements follow from manipulating with the identity (7.1.7) in a similar way as the computation (7.1.8) in Example 7.1.6. For the third statement, we claim that the T-equivariant version is $\mathbf{t}^{e_{[n]}} KT^{T^{-1}}_{\mathcal{F}}(y,x) = KT_{\mathcal{F}^\vee}(x,y)$ (where the T^{-1} superscript means that we have replaced t_i by t_i^{-1}). Verifying this identity is then another easy manipulation with (7.1.7). $\qquad\square$

7.2 Summations of lattice point generating functions

Here we collect some useful results concerning summations of lattice point generating functions arising from polyhedra, along with slight variants that are suitable for our purposes. The key two results are Theorem 7.2.8 and Theorem 7.2.12. We point to [BHS09] and [Pos09, §19 (Appendix)] as helpful references.

7.2.1 Brion's formula

Here we review the results in [Bri88, Ish90]. For a subset $S \subset \mathbb{R}^n$, denote by $\mathbb{1}(S) : \mathbb{Z}^n \to \mathbb{Q}$ its indicator function sending $x \mapsto 1$ if $x \in S$ and 0 otherwise. Let $\mathcal{P}_n$ be the vector space of $\mathbb{Q}$-valued functions on $\mathbb{Z}^n$ generated by $\{\mathbb{1}(P) \mid P \subset \mathbb{R}^n \text{ lattice polyhedron}\}$. It follows from the Brianchon-Gram formula [Bri37, Gra74, She67] that $\mathcal{P}_n$ is generated by indicator functions of cones, and by triangulating one concludes that $\mathcal{P}_n$ is generated by indicator functions of smooth cones. We will often consider elements of $\mathcal{P}_n$ as elements of the power series ring $\mathbb{Q}[[t_1^{\pm}, \ldots, t_n^{\pm}]]$ by identifying $\mathbb{1}(P)$ with $\sum_{\lambda \in P \cap \mathbb{Z}^n} \mathbf{t}^{\lambda}$. The following fundamental theorem concerns convergence of these power series to a rational function.

Theorem 7.2.1 ([Ish90, Theorem 1.2]). *Consider $\mathcal{P}_n$ as $\mathbb{Q}[t_1^{\pm}, \ldots, t_n^{\pm}]$-submodule of $\mathbb{Q}[[t_1^{\pm}, \ldots, t_n^{\pm}]]$, and let $\mathbb{Q}(t_1, \ldots, t_n)$ be the field of rational functions. There exists a unique $\mathbb{Q}[t_1^{\pm}, \ldots, t_n^{\pm}]$-linear map*

$$\text{Hilb} : \mathcal{P}_n \to \mathbb{Q}(t_1, \ldots, t_n)$$

such that if $C \subset \mathbb{R}^n$ is a smooth cone with ray generators $\mathbf{u}_1, \ldots, \mathbf{u}_k \in \mathbb{Z}^n$ and vertex at the origin then $\text{Hilb}(\mathbb{1}(C)) = \text{Hilb}(C) = \prod_{i=1}^{k} \frac{1}{1 - \mathbf{t}^{\mathbf{u}_i}}$.

Two remarks about the above linear map Hilb follow:

1. The notation Hilb agrees with our previous notion of Hilbert series from Section 6.2.1: when C is a pointed rational polyhedral cone, not necessarily smooth, $\text{Hilb}(\mathbb{1}(C)) = \text{Hilb}(C)$.

2. For P a lattice polyhedron with non-trivial lineality space, $\text{Hilb}(\mathbb{1}(P)) = 0$.

For P a lattice polyhedron, we will often by abuse of notation write $\text{Hilb}(P)$ for $\text{Hilb}(\mathbb{1}(P))$. An important result on rational generating functions for cones is Brion's formula [Bri88], which was extended to a slightly more general version in [Ish90]. We will only need the following special case of [Ish90, Theorem 2.3].

Theorem 7.2.2. *Let $P \subset \mathbb{R}^n$ be a lattice polyhedron with a nonempty set of vertices (so P has no lineality space), and let $C(P)$ be its recession cone. For every vertex $\mathbf{v}$ of P, write $C_{\mathbf{v}}$ for $\text{Cone}(P - \mathbf{v})$. Then we have*

$$\text{Hilb}(P) = \sum_{\mathbf{v} \in \text{Vert}(P)} \text{Hilb}(C_{\mathbf{v}} + \mathbf{v}) \qquad \text{and} \qquad \text{Hilb}(C(P)) = \sum_{\mathbf{v} \in \text{Vert}(P)} \text{Hilb}(C_{\mathbf{v}}).$$

Proposition 7.2.3. *For $\mathcal{F}$ a flag matroid on $[n]$, it holds that*

$$K\mathcal{T}_{\mathcal{F}}^{T}(1, 1) = \text{Hilb}(P(\mathcal{F})).$$

Proof. This follows immediately from (7.1.7) and Brion's formula. $\square$

7.2.2 Flipping cones

Here we review the method of flipping cones [FS10, §6], [BHS09, (11)]. Our contribution is a generalization (Theorem 7.2.8), which will serve as a key technical tool in the next section. Let $\zeta \in \mathbb{R}^n$. For every $a \in \mathbb{R}$, we will denote the hyperplane $\{x \in \mathbb{R}^n \mid \langle \zeta, \mathbf{x} \rangle = a\}$ by $H_{\zeta=a}$ and the half-space $\{x \in \mathbb{R}^n \mid \langle \zeta, \mathbf{x} \rangle \geq a\}$ by $H_{\zeta \geq a}$. For an element $f \in \mathcal{P}_n$, by considering f as an element of $\mathbb{Q}[[t_1^{\pm}, \ldots, t_n^{\pm}]]$ we write $f|_{H_{\zeta=a}}$ for the sum of terms $c\mathbf{t}^{\mathbf{w}}$ in f such that $\langle \zeta, \mathbf{w} \rangle = a$.

Definition 7.2.4. A polyhedron $P \subset \mathbb{R}^n$ is ζ-*pointed* if $P \subseteq H_{\zeta \geq a}$ for some $a \in \mathbb{R}$. Let $\mathcal{P}_n^\zeta$ be the vector space of ζ-pointed elements in $\mathcal{P}_n$.

We note the following useful observation: Let $P \subset \mathbb{R}^n$ be a polyhedron with vertices $\mathrm{Vert}(P)$, and as before let $C_\mathbf{v} := \mathrm{Cone}(P - \mathbf{v})$ for $\mathbf{v} \in \mathrm{Vert}(P)$. For $\zeta \in \mathbb{R}^n$, the cone $C_\mathbf{v}$ is ζ-pointed if and only if $\mathbf{v}$ is a vertex of the face $P^{-\zeta}$ of P minimizing the linear function $\langle \zeta, . \rangle$ on P. If $f \in \mathcal{P}_n^\zeta$, then one can compute $\mathrm{Hilb}(f)$ "slice-by-slice" in the following sense.

Lemma 7.2.5. *Let $f, g \in \mathcal{P}_n^\zeta$ and suppose that $\mathrm{Hilb}(f) = \mathrm{Hilb}(g)$. Then for every $a \in \mathbb{R}$, it holds that $\mathrm{Hilb}(f|_{H_{\zeta=a}}) = \mathrm{Hilb}(g|_{H_{\zeta=a}})$.*

Proof. Write $b = f - g$, and suppose by contradiction that there is an $a \in \mathbb{R}$ with $\mathrm{Hilb}(b|_{H_{\zeta=a}}) \neq 0$. Since $b \in \mathcal{P}_n^\zeta$, there is a minimal such a, which we will denote by a_0. By the *Claim* below, we can find a nonzero $q = \sum_{e \in \mathbb{Z}^n} \lambda_e t^e \in \mathbb{Q}[t_1^{\pm}, \ldots, t_n^{\pm}]$ such that $q \cdot b$ has finite support, i.e. is a Laurent polynomial. So $\mathrm{Hilb}(q \cdot b) = q\,\mathrm{Hilb}(b) = 0$. Since $q \cdot b$ has finite support, this implies that $q \cdot b = 0$. Let $c = \min\{\langle \zeta, e \rangle \mid \lambda_e \neq 0\}$, and let $q_0 = \sum_{e:\langle \zeta, e \rangle = c} \lambda_e t^e$. Then $0 = \mathrm{Hilb}((q \cdot b)|_{H_{\zeta=a_0+c}}) = q_0 \,\mathrm{Hilb}(b|_{H_{\zeta=a_0}}) \neq 0$, a contradiction.

Claim: For every $f \in \mathcal{P}_n$, there exists a nonzero Laurent polynomial $q \in \mathbb{Q}[t_1^{\pm}, \ldots, t_n^{\pm}]$ such that $q \cdot f$, which is a priori an element of the power series ring $\mathbb{Q}[[t_1^{\pm}, \ldots, t_n^{\pm}]]$, is a Laurent polynomial in $\mathbb{Q}[t_1^{\pm}, \ldots, t_n^{\pm}]$.

Proof of claim: If $f = \mathbb{1}(C)$ for some smooth cone C, one can take $q(\mathbf{t}) = \prod (1 - \mathbf{t}^e)$, where the product is over the primitive ray generators of C. Since $\mathcal{P}_n$ is generated by smooth cones, the result follows. $\square$

Suppose that ζ is chosen such that the ζ_i are $\mathbb{Q}$-linearly independent (in the future, we will abbreviate this to "ζ is *irrational*"). Then for every $a \in \mathbb{R}$, the intersection $H_{\zeta=a} \cap \mathbb{Z}^n$ consists of at most one point. In this case Lemma 7.2.5 reduces to saying that $\mathrm{Hilb} : \mathcal{P}_n^\zeta \to \mathbb{Q}(t_1, \ldots, t_n)$ is injective, and we recover [FS10, Lemma 6.3]. We next recall the notion of cone flips. We begin with a lemma for their existence.

Lemma 7.2.6 ([Haa05, Lemma 6], [FS12, Lemma 2.1]). *Assume ζ is irrational. For every $f \in \mathcal{P}_n$, there is a unique $f^\zeta \in \mathcal{P}_n^\zeta$ such that $\mathrm{Hilb}(f) = \mathrm{Hilb}(f^\zeta)$. The map $f \mapsto f^\zeta$ is linear.*

The map $(\cdot)^\varsigma$ in the lemma above can be described explicitly as follows. Let $C \subseteq \mathbb{R}^n$ be the support function of a rational simplicial cone, i.e.

$$C = \mathbb{1}\left(\left\{\mathbf{v} + \sum_{i=0}^{n-1} a_i \mathbf{u}_i \mid a_i \geq 0 \text{ for all } i \in [n]\right\}\right).$$

Then the image $C^\varsigma \in \mathcal{P}_n^\varsigma$ under the map of Lemma 7.2.6 is given by

$$C^\varsigma = (-1)^\ell \mathbb{1}\left(\left\{\mathbf{v} + \sum_{i=0}^{n-1} a_i \mathbf{u}_i \,\middle|\, \begin{array}{l} a_i \geq 0 \text{ for all } i \text{ with } \langle \boldsymbol{\zeta}, \mathbf{u}_i \rangle > 0, \\ \text{and } a_i < 0 \text{ for all } i \text{ with } \langle \boldsymbol{\zeta}, \mathbf{u}_i \rangle < 0 \end{array}\right\}\right), \quad (7.2.1)$$

where ℓ is the number of rays $\mathbf{u}_i$ for which $\langle \boldsymbol{\zeta}, \mathbf{u}_i \rangle < 0$. This is a consequence of the fact that $\mathrm{Hilb}(\mathbb{1}(P)) = 0$ if P has a nontrivial lineality space. We will refer to C^ς as the *cone flip* of C in direction $\boldsymbol{\zeta}$. For a non-simplicial rational cone C, one defines the flipped cone $C^\varsigma \in \mathcal{P}_n^\varsigma$ by triangulating the cone [1].

Remark 7.2.7. The assumption that $\boldsymbol{\zeta}$ is irrational is essential for Lemma 7.2.6: if $\boldsymbol{\zeta}$ is not irrational then $\mathcal{P}_n^\varsigma$ contains some lattice polyhedron P with a non-trivial lineality space, and $\mathrm{Hilb}(P) = 0 = \mathrm{Hilb}(0)$, contradicting uniqueness.

Now, suppose we are given an expression

$$\varphi = \sum_{\lambda \in \Lambda} a_\lambda \mathrm{Hilb}(C_\lambda) \in \mathbb{Q}(t_1, \ldots, t_n), \quad (7.2.2)$$

which is a finite summation where the C_λ are pointed cones with vertices not necessarily at the origin and $a_\lambda \in \mathbb{Q}$ are scalars. Suppose we know that $\varphi \in \mathbb{Q}(t_1, \ldots, t_n)$ is in fact a Laurent polynomial (for example, because φ arose from a computation in T-equivariant K-theory). Then we can use cone-flipping to get partial information about the coefficients of φ. The following proposition is our "cone-flipping in slices" technique which will be repeatedly used in later sections.

Theorem 7.2.8. *Suppose* $\varphi = \sum_\lambda a_\lambda \mathrm{Hilb}(C_\lambda)$ *is a Laurent polynomial, i.e.* $\varphi \in \mathbb{Q}[t_1^\pm, \ldots, t_n^\pm]$*, and let* P *be the convex hull of the vertices of the* C_λ*. For* $\boldsymbol{\zeta} \in \mathbb{R}^n$*, not necessarily irrational, assume that every cone* C_λ *whose vertex* $\mathbf{v}_\lambda$ *satisfies* $\langle \boldsymbol{\zeta}, \mathbf{v}_\lambda \rangle < b$ *is* $\boldsymbol{\zeta}$*-pointed. Then*

$$\varphi|_{H_{\zeta=b}} = \sum_{C_\lambda \in \mathcal{P}_n^\varsigma} a_\lambda \mathrm{Hilb}(C_\lambda \cap H_{\zeta=b}).$$

Note that if $P \cap H_{\zeta=b}$ is the face $P^{-\varsigma}$ of P minimizing in the $\boldsymbol{\zeta}$-direction, the assumption in the proposition is always satisfied, and all cones contributing to the right hand side have their vertex on $P^{-\varsigma}$.

---

[1] We remark that calling C^ς the "flipped cone" of C is a slight abuse of terminology when C is not simplicial, since C^ς is not necessarily the support function of a polyhedron up to sign. It can be a genuine linear combination of some of those; see [FS10, Remark 6.7].

Remark 7.2.9. In the special case of Theorem 7.2.8 where $H_{\zeta=b} \cap P = \{\mathbf{v}\}$ is a vertex of P, the coefficient of $\mathbf{t}^{\mathbf{v}}$ in φ is equal to $\sum a_i$, where the sum is over all λ for which $C_\lambda \in \mathcal{P}_n^\zeta$ and the vertex of C_λ is at $\mathbf{v}$. Moreover, if $\mathbf{w}$ is a vertex of the Newton polytope $\mathrm{Newt}(\varphi)$ of φ, then for any irrational $\zeta \in \mathbb{R}^n$ there must exist a cone C_λ such that its vertex $\mathbf{v}_\lambda$ satisfies $\langle \zeta, \mathbf{w} \rangle \leq \langle \zeta, \mathbf{v}_\lambda \rangle$. In other words, Theorem 7.2.8 is a generalization of [FS10, Corollary 6.9], which states that the Newton polytope of φ is contained in P.

We prepare for the proof by noting a useful feature of the cone-flipping operation, starting with the following notion.

Definition 7.2.10. Let C be a pointed cone, and $\zeta \in \mathbb{R}^n$. We will say that an irrational $\zeta' \in \mathbb{R}^n$ is an *irrational approximation* of ζ with respect to C, if for every ray generator $\mathbf{u} \in \mathbb{R}^n$ of C it holds that $\langle \zeta, \mathbf{u} \rangle > 0 \implies \langle \zeta', \mathbf{u} \rangle > 0$ and $\langle \zeta, \mathbf{u} \rangle < 0 \implies \langle \zeta', \mathbf{u} \rangle < 0$.

Note that an irrational approximation of ζ can always be obtained as a small perturbation of ζ. The following is a minor generalization of [FS12, Lemma 2.3], with almost identical proof, which we have included for completeness.

Lemma 7.2.11. *Let $\zeta \in \mathbb{R}^n$, let C be a pointed cone with vertex at $\mathbf{v}$, and let $\zeta' \in \mathbb{R}^n$ be an irrational approximation of ζ. Then $C^{\zeta'}$ is supported in the half space $\{\mathbf{x} \mid \langle \zeta, \mathbf{x} \rangle \geq \langle \zeta, \mathbf{v} \rangle\}$. Furthermore, if C is not contained in $\{\mathbf{x} \mid \langle \zeta, \mathbf{x} \rangle \geq \langle \zeta, \mathbf{v} \rangle\}$, then $C^{\zeta'}$ is supported in the open half space $\{\mathbf{x} \mid \langle \zeta, \mathbf{v} \rangle > \langle \zeta, \mathbf{v} \rangle\}$; in particular $w \notin C^{\zeta'}$.*

Proof. If C is simplicial, the result follows immediately from the construction of cone flips (7.2.1) and Definition 7.2.10. For general C, we can obtain the first statement by considering any triangulation of C. For the second one, choose a ray $\mathbf{u}$ of C such that $\langle \zeta, \mathbf{u} \rangle < 0$ and a triangulation of C such that every interior cone contains $\mathbf{u}$. Such a triangulation can for instance be constructed by triangulating the faces of C that do not contain $\mathbf{u}$, and then coning that triangulation from $\mathbf{u}$. Now $C = \sum_F (-1)^{\dim C - \dim F} \mathbb{1}(F)$ and $C^{\zeta'} = \sum_F (-1)^{\dim C - \dim F} \mathbb{1}(F)^{\zeta'}$, where the sum is over all interior cones of the triangulation. The result now follows from the simplicial case. $\qquad\square$

Proof of Theorem 7.2.8. Since the summation defining φ is over a finite collection of cones $\{C_\lambda\}_{\lambda \in \Lambda}$, there exists a $\zeta' \in \mathbb{R}$ which is an irrational approximation of ζ with respect to every cone C_λ. By assumption $\varphi = \mathrm{Hilb}(f)$, where $f \in \mathcal{P}_n$ has finite support, in particular $f \in \mathcal{P}_n^\zeta$. Hence, by Lemma 7.2.5, $\varphi|_{H_{\zeta=b}} = \mathrm{Hilb}(\sum a_\lambda \mathbb{1}(C_\lambda^{\zeta'} \cap H_{\zeta=b}))$. If $C_\lambda \notin \mathcal{P}_n^\zeta$, then by assumption the vertex $\mathbf{v}_\lambda$ of C_λ satisfies $\langle \zeta, \mathbf{v}_\lambda \rangle \geq b$, and by Lemma 7.2.11 $C_i^{\zeta'}$ is supported on the open half-space $\{x \mid \langle \zeta, x \rangle > b\}$, in particular $C_i^{\zeta'} \cap H_{\zeta=b} = \emptyset$. If $C_\lambda \in \mathcal{P}_n^\zeta$, then since C_i and $C_\lambda^{\zeta'}$ are both in $\mathcal{P}_n^\zeta$, it follows from Lemma 7.2.5 that $\mathrm{Hilb}(C_\lambda^{\zeta'} \cap H_{\zeta=b}) = \mathrm{Hilb}(C_\lambda \cap H_{\zeta=b})$. $\qquad\square$

7.2.3 Flipping cones for base polytopes

Let us now specialize our discussion of summing lattice point generating functions to ones arising from flag varieties. For the rest of this section, let $\mathcal{F}$ be a flag matroid of rank $\mathbf{r} = (r_1, \ldots, r_k)$ on a ground set $[n]$. As before, for a basis F of $\mathcal{F}$ let us write $\mathrm{Cone}_F(\mathcal{F}) := \mathrm{Cone}(P(\mathcal{F}) - e_F))$ and $\mathrm{Hilb}_F(\mathcal{F}) := \mathrm{Hilb}(\mathrm{Cone}_F(\mathcal{F}))$.

Consider the expression below, which is a finite summation

$$\varphi = \sum_{\lambda \in \Lambda} a_\lambda \mathbf{t}^{\mathbf{w}_\lambda} \mathrm{Hilb}_{F^\lambda}(\mathcal{F}), \tag{7.2.3}$$

where $a_\lambda \in \mathbb{Q}$, $\mathbf{w}_\lambda \in \mathbb{Z}^n$, and F^λ a basis of $\mathcal{F}$. We allow the same basis to occur several times in the sum. Note that $\mathbf{t}^{\mathbf{w}_\lambda} \mathrm{Hilb}_{F^\lambda}(\mathcal{F}) = \mathrm{Hilb}(C_\lambda)$, where C_λ is a cone with vertex at $\mathbf{w}_\lambda$, so (7.2.3) is a special case of (7.2.2). As before, we assume that $\varphi \in \mathbb{Q}[t_1^\pm, \ldots, t_n^\pm]$, i.e. φ is a Laurent polynomial, and we write $P := \mathrm{Conv}(\mathbf{w}_\lambda \mid \lambda \in \Lambda)$ for the convex hull of the $\mathbf{w}_\lambda$. We will assume that all $\mathbf{w}_\lambda$ lie in $\mathbb{Z}_{\geq 0}^n$, and that there exists a $c \in \mathbb{Z}_{\geq 0}$ such that the sum of the entries of any $\mathbf{w}_\lambda$ is equal to c. Let $\widetilde{P} := \mathrm{Conv}(\sigma \cdot \mathbf{w}_\lambda \mid \sigma \in S_n, \ \lambda \in \Lambda)$ be the convex hull of all points in $\mathbb{Z}_{\geq 0}^n$ that are equal to one of the $\mathbf{w}_i$ up to permuting entries. The following theorem will be repeatedly applied in the next section.

Theorem 7.2.12. *Let φ and $\widetilde{P}$ be as above, and let $\mathbf{v}$ be a vertex of $\widetilde{P}$. Write $\mathbf{v} = e_{S_1} + \cdots + e_{S_m}$, with $S_1 \subseteq \ldots \subseteq S_m \subseteq [n]$. Fix a basis $F = (B_1, \ldots, B_k)$ of $\mathcal{F}$ such that e_F is a vertex of the face $P(\mathcal{F})^{\mathbf{v}}$ of $P(\mathcal{F})$ maximizing the direction $\mathbf{v}$, that is, a basis F satisfying $|S_i \cap B_j| = r_{M_j}(S_i)$ for all $1 \leq i \leq m$ and $1 \leq j \leq k$ (Proposition 5.3.17). Then the coefficient of $\mathbf{t}^{\mathbf{v}}$ in $\varphi \in \mathbb{Q}[t_1^\pm, \ldots, t_n^\pm]$ is equal to the sum of all a_λ for which $\mathbf{w}_\lambda = \mathbf{v}$ and $F^\lambda = F$.*

Proof. If $\mathbf{v} \notin P$, the result follows from Remark 7.2.9. So we consider the case $\mathbf{v} \in P$. Let us write $\mathbf{v} = (v_1, \ldots, v_n)$ and $e_F = (b_1, \ldots, b_n)$. By permuting the coordinates of $\mathbb{N}^n$, let us assume that $v_i \geq v_{i+1}$ for all $i \in [n]$, and that $b_i \geq b_{i+1}$ whenever $v_i = v_{i+1}$. Let

$$\zeta' := ne_1 + (n-1)e_2 + \cdots + 2e_{n-1} + e_n.$$

We claim that this ζ' has the following properties.

1. The vertex $\{\mathbf{v}\}$ is the face of $\widetilde{P}$ maximizing in the ζ' direction, and hence is the vertex of P maximizing in the ζ' direction.

2. The vertex face of $P(\mathcal{F})$ maximizing in the ζ' direction is $\{e_F\}$.

The first property is immediate from the construction of ζ', as we have assumed that $v_i \geq v_{i+1}$ for all $i = 1, \ldots, n$. For the second property, note that ζ' is an interior point in the cone

$$\mathrm{Cone}(e_1, e_1 + e_2, \ldots, e_1 + \cdots + e_{n-1}) + \mathbb{R}e_{[n]},$$

of which the cone

$$\mathrm{Cone}(e_{S_1}, e_{S_2}, \ldots, e_{S_m}) + \mathbb{R}e_{[n]}$$

is a face. This face contains $\mathbf{v}$ in its relative interior. These two cones are cones in the braid arrangement, of which the normal fan of $P(\mathcal{F})$ is a coarsening (Theorem 5.3.15). Thus, the vertex face of $P(\mathcal{F})$ maximizing in the ζ' direction is among the vertices of $P(\mathcal{F})^{\mathbf{v}}$, and our assumption $b_i \geq b_{i+1}$ for all $i = 1, \ldots, n$ such that $v_i = v_{i+1}$ ensures that e_F is indeed the one. Now, applying Theorem 7.2.8 (in the form of Remark 7.2.9) with $\zeta = -\zeta'$ gives the desired statement. $\qquad\square$

7.3 Combinatorial properties for 2-step flag matroids

7.3.1 Some terms of the flag-geometric Tutte polynomial

As a first application of the methods from Section 7.2, we apply Theorem 7.2.12 to compute some of the terms in the expression (7.1.7) for $K\mathcal{T}_{\mathcal{F}}^T(u+1, v+1)$.

Theorem 7.3.1. *Let $\mathcal{F} = (M_1, M_2)$ be a 2-step flag matroid and let $\mathbf{t}^{\mathbf{k}} u^{r_2-i} v^j$ be a monomial occurring in (7.1.7). Then $\sum k_\ell = r_1 + i + j$. Let c denote the number of entries in $\mathbf{k}$ that are equal to 1. If $c \leq |r_1 + j - i|$, the coefficient of $\mathbf{t}^{\mathbf{k}} u^{r_2-i} v^j$ in $K\mathcal{T}_{\mathcal{F}}^T(u+1, v+1)$ is equal to*

1. 1, if S_2 is spanning for M_1, S_1 is independent in M_2, and $c = |r_1 + j - i|$,

2. 0, otherwise,

where S_1 and S_2 are defined by $S_1 \subseteq S_2$ and $\mathbf{k} = e_{S_1} + e_{S_2}$.

Proof. The equality $\sum k_\ell = r_1 + i + j$ follows immediately from (7.1.7). The coefficient of $\mathbf{t}^{\mathbf{k}} u^{r_2-i} v^j$ is equal to

$$\sum_{\substack{F = (B_1, B_2), \\ F \in \mathcal{B}(\mathcal{F})}} \mathrm{Hilb}_F(\mathcal{F}) \sum_{\substack{\mathfrak{p} \subseteq B_2, \, \mathfrak{q} \subseteq J_1, \\ |\mathfrak{p}| = i \quad |\mathfrak{q}| = j}} t^{e_{B_1} + e_{\mathfrak{p}} + e_{\mathfrak{q}}}. \tag{7.3.1}$$

where we have denoted $J_1 := [n] \setminus B_1$. It is not hard to see that the vertices of $\tilde{P}$ have $|r_1 + j - i|$ entries equal to 1. This proves that the coefficient is 0 if $c < |r_1 + j - i|$. So from now on we assume $c = |r_1 + j - i|$.

Next, we apply Theorem 7.2.12. Writing $\mathbf{k} = e_{S_1} + e_{S_2}$, (note that $|S_1| = \min(i, r_1 + j)$ and $|S_2| = \max(i, r_1 + j)$) we find a basis of $\mathcal{F}$ for which $r_j(S_i) = |S_i \cap B_j|$. We now need to compute the number of ways $\mathbf{k}$ can be written as a sum $e_{B_1} + e_{\mathfrak{p}} + e_{\mathfrak{q}}$. If S_2 is not spanning for M_1, or if S_1 is not independent in M_2, there are no ways to do this, and the coefficient is 0. Otherwise, if $i \leq r_1 + j$, we need to put $\mathfrak{p} = S_1$ and $\mathfrak{q} = S_2 \setminus S_1$. If $i \geq r_1 + j$, we need to put $\mathfrak{q} = S_1 \cap J_1$ and $\mathfrak{p} = S_1 \cup J_1$. In both cases, there is just one way, so the coefficient is 1. $\qquad\square$

7.3.2 Towards a corank-nullity formula

For a matroid M on $[n]$, the corank-nullity formula for the Tutte polynomial $T_M(x, y) = \sum_{S \subseteq [n]} (x-1)^{r-r(S)} (y-1)^{|S|-r(S)}$ expresses T_M as a sum over all subsets of $[n]$. In particular, we have $T_M(2, 2) = 2^n$; in fact, $KT_M^T(2, 2) = \prod_{i=1}^{n} (1+t_i)$. As a first step towards a corank-nullity-type formula for $KT_{\mathcal{F}}$, we show the following for a two-step flag matroid.

Theorem 7.3.2. *Let $\mathcal{F}$ be a two-step flag matroid $\mathcal{F} = (M_1, M_2)$ of rank (r_1, r_2). Let $p\mathcal{B}(\mathcal{F})$ be the set of pseudo-bases of $\mathcal{F}$, i.e. subsets $S \subseteq [n]$ such that S is spanning in M_1 and independent in M_2. With q as a formal variable, we have*

$$KT_{\mathcal{F}}^T(1+q^{-1}, 1+q) = q^{-r_2}\Big(\prod_{i=1}^{n}(1+t_i q)\Big)\Big(\sum_{S \in p\mathcal{B}(\mathcal{F})} \mathbf{t}^{e_S} q^{|S|}\Big),$$

and in particular, we have

$$KT_{\mathcal{F}}(1+q^{-1}, 1+q) = q^{-r_2} \cdot 2^n \cdot \Big(\sum_{S \in p\mathcal{B}(\mathcal{F})} q^{|S|}\Big),$$

$$KT_{\mathcal{F}}^T(2, 2) = \Big(\prod_{i=1}^{n}(1+t_i)\Big)\Big(\sum_{S \in p\mathcal{B}(\mathcal{F})} \mathbf{t}^{e_S}\Big), \text{ and}$$

$$KT_{\mathcal{F}}(2, 2) = 2^n |p\mathcal{B}(\mathcal{F})|.$$

Proof. Setting $u = q^{-1}$ and $v = q$ in (7.1.7) of Theorem 7.1.4 gives us

$$KT_{\mathcal{F}}^T(1+q^{-1}, 1+q) = \sum_{\substack{F = (B_1, B_2), \\ F \in \mathcal{B}(\mathcal{F})}} \mathrm{Hilb}_F(\mathcal{F}) \sum_{\mathfrak{p} \subseteq B_2} \sum_{\mathfrak{q} \subseteq [n] \setminus B_1} \mathbf{t}^{e_{B_1} + e_{\mathfrak{p}} + e_{\mathfrak{q}}} q^{|\mathfrak{p}| + |\mathfrak{q}| - r_2}$$

$$= \sum_{\substack{F = (B_1, B_2), \\ F \in \mathcal{B}(\mathcal{F})}} \mathrm{Hilb}_F(\mathcal{F}) \sum_{R \subseteq [n]} \sum_{S \subseteq B_2 \setminus B_1} \mathbf{t}^{e_{B_1} + e_R + e_S} q^{|R| + |S| - r_2}$$

$$= q^{-r_2} \prod_{i=1}^{n}(1+t_i q) \sum_{\substack{F = (B_1, B_2), \\ F \in \mathcal{B}(\mathcal{F})}} \mathrm{Hilb}_F(\mathcal{F}) \sum_{S \subseteq B_2 \setminus B_1} \mathbf{t}^{e_{B_1} + e_S} q^{|S|}$$

We now use Theorem 7.2.12 to compute the sum

$$\varphi_r := \sum_{\substack{F = (B_1, B_2), \\ F \in \mathcal{B}(\mathcal{F})}} \mathrm{Hilb}(\mathrm{Cone}_F(\mathcal{F})) \sum_{\substack{B_1 \subseteq \mathfrak{p} \subseteq B_2, \\ |\mathfrak{p}| = r}} t^{e_{\mathfrak{p}}}$$

for a fixed $r_1 \leq r \leq r_2$. First, we note that the polytope $\widetilde{P} = \mathrm{Conv}(e_S \mid S \subseteq E, |S| = r)$, obtained as the convex hull of the S_n-orbit of $\{e_{\mathfrak{p}} \mid B_1 \subseteq \mathfrak{p} \subseteq B_2, |\mathfrak{p}| = r\}$, has no interior lattice points.

For $S \subseteq E$ with $|S| = r$, if S is not a pseudo-basis of $\mathcal{F}$, then there is no basis F of $\mathcal{F}$ such that $B_1 \subseteq S \subseteq B_2$, and hence the coefficient of $\mathbf{t}^{es}$ is 0 in this case. Now, suppose S is a pseudo-basis of $\mathcal{F}$, which by definition implies that there exists a basis $F = (B_1, B_2)$ of $\mathcal{F}$ with $B_1 \subseteq S \subseteq B_2$. This basis F is a vertex of the face $P(\mathcal{F})^{es}$ by Proposition 5.3.17, and thus by Theorem 7.2.12 the coefficient of $\mathbf{t}^{es}$ is equal to 1 in φ_r. $\qquad\square$

We do not know of analogues of Theorem 7.3.2 for flag matroids with more than two constituents.

7.3.3 A deletion-contraction-like relation

In this section, we consider $K\mathcal{T}_{\mathcal{F}}$ of an elementary quotient $\mathcal{F} = (M_1, M_2)$ (i.e. $r(M_2) - r(M_1) = 1$). In this case, there is a unique matroid M on a ground set $[\tilde{n}] := \{0\} \sqcup [n]$ such that $M_1 = M/0$ and $M_2 = M \backslash 0$. Our main theorem of this subsection is the following deletion-contraction-like relation.

Theorem 7.3.3. *Let M be a matroid of rank r on $[\tilde{n}] := \{0\} \sqcup [n]$ such that the element 0 is neither a loop or a coloop in M. Let $\tilde{T} = \mathbb{C}^* \times T = (\mathbb{C}^*)^{n+1}$ be the torus with character ring $\mathbb{Z}[t_0^{\pm}, \ldots, t_n^{\pm}]$. Then we have*

$$K\mathcal{T}_{M,M}^{\tilde{T}}(x, y) = t_0^2 K\mathcal{T}_{M/0, M/0}^{T}(x, y) + t_0 K\mathcal{T}_{M/0, M\backslash 0}^{T}(x, y) + K\mathcal{T}_{M\backslash 0, M\backslash 0}^{T}(x, y). \quad (7.3.2)$$

In particular, we have

$$K\mathcal{T}_{M,M}(x, y) = K\mathcal{T}_{M/0, M/0}(x, y) + K\mathcal{T}_{M/0, M\backslash 0}(x, y) + K\mathcal{T}_{M\backslash 0, M\backslash 0}(x, y).$$

We use $\{e_0, \ldots, e_n\}$ for the standard basis of $\mathbb{R}^{n+1} = \mathbb{R} \oplus \mathbb{R}^n$. For a polyhedron $P \subset \mathbb{R}^n$, we will often abuse the notation and also write P also for $\{0\} \times P \subset \mathbb{R} \oplus \mathbb{R}^n$. We prepare for the proof of Theorem 7.3.3 by an observation that motivated the theorem.

As the base polytope $P(M)$ is a $(0, 1)$-polytope (i.e. a lattice polytope contained in the Boolean cube $[0, 1]^{n+1} \subset \mathbb{R}^{n+1}$), every lattice point is a vertex. Moreover, observe that the vertices of $P(M)$ partition into two parts, the bases of $M/0$ and the bases of $M\backslash 0$. As a result, the lattice points of $P(M, M) = P(M) + P(M)$ partition into the following three parts, with $P_1 = \frac{1}{2}(P_0 + P_2)$:

- $P_2 := P(M, M) \cap H_{e_0 = 2} = \{2e_0\} \times P(M/0, M/0)$,

- $P_1 := P(M, M) \cap H_{e_0 = 1} = \{e_0\} \times P(M/0, M\backslash 0)$, and

- $P_0 := P(M, M) \cap H_{e_0 = 0} = \{0\} \times P(M\backslash 0, M\backslash 0)$.

The case of setting $x = y = 1$ (cf. Proposition 7.1.7.(4)) in (7.3.2) of Theorem 7.3.3 witnesses this partition of the lattice points of $P(M, M)$. The following lemma in preparation for the proof of Theorem 7.3.3 is a consequence of $P_1 = \frac{1}{2}(P_0 + P_2)$.

Lemma 7.3.4. *Let the notations be as above. Then for $B \in \mathcal{B}(M)$ with $0 \notin B$, we have*

$$\mathrm{Hilb}(\mathrm{Cone}_B(P(M,M)) \cap H_{e_0=1}) = \sum_{\substack{I \in \mathcal{B}(M/0), \\ I \subset B}} t_0 t_{B\backslash I}^{-1} \, \mathrm{Hilb}_{(I,B)}(M/0, M\backslash 0) \text{ and}$$

$$\mathrm{Hilb}(\mathrm{Cone}_B(P(M,M)) \cap H_{e_0=0}) = \sum_{\substack{I \in \mathcal{B}(M/0), \\ I \subset B}} \mathrm{Hilb}_{(I,B)}(M/0, M\backslash 0).$$

Proof. We have an equality of polyhedra

$$\mathrm{Cone}_B(P(M,M)) \cap H_{e_0=1} = \mathrm{Cone}_B(P(M\backslash 0)) + P_1 - 2e_B.$$

We claim that $\mathrm{Cone}_B(P(M\backslash 0)) + P_1$ has vertices $\{e_I + e_B\}$ for $I \in \mathcal{B}(M/0)$ such that $I \subset B$. The two statements in the lemma then follow from Brion's formula Theorem 7.2.2. For the claim, we start by noting that if $I \in \mathcal{B}(M/0)$ then there exists $B' \in \mathcal{B}(M\backslash 0)$ such that $I \subset B'$ (since $M/0 \twoheadleftarrow M\backslash 0$). Consequently, if e_B is the vertex of $P(M\backslash 0)$ that minimizes $\langle \mathbf{v}, e_B \rangle$ for some $\mathbf{v} \in \mathbb{R}^n$, then a vertex of $P(M/0)$ that minimizes $\langle \mathbf{v}, \cdot \rangle$ must be e_I satisfying $I \subset B$. Our claim now follows from $P_1 = \frac{1}{2}(P_0 + P_2)$. $\qquad\square$

Proof of Theorem 7.3.3. Let us begin by noting that the equation (7.1.7) for $K\mathcal{T}_{M,M}^{\widetilde{T}}$ reads

$$K\mathcal{T}_{M,M}^{\widetilde{T}}(u+1, v+1) = \sum_{B \in \mathcal{B}(M)} \mathrm{Hilb}_B(P(M,M)) \sum_{\mathfrak{p} \subseteq B} \sum_{\mathfrak{q} \subseteq [\widetilde{n}] \backslash B} \mathbf{t}^{e_B + e_{\mathfrak{p}} + e_{\mathfrak{q}}} u^{r-|\mathfrak{p}|} v^{|\mathfrak{q}|}.$$

$$(7.3.3)$$

We apply Theorem 7.2.8 with $\zeta = e_0$ and L defined by $t_0 = 0$. Note that $\mathrm{Cone}_B(P(M,M)) \in \mathcal{P}_n^\zeta$ if and only if $0 \notin B$. Hence all cones occurring in (7.3.3) with vertex on L are in $\mathcal{P}_n^\zeta$, and we find that the terms in (7.3.3) not divisible by t_0 sum to

$$\sum_{\substack{B \in \mathcal{B}(M), \\ 0 \notin B}} \mathrm{Hilb}(\mathrm{Cone}_B(P(M,M)) \cap H_{e_0=0}) \sum_{\mathfrak{p} \subseteq B} \sum_{\mathfrak{q} \subseteq [n] \backslash B} \mathbf{t}^{e_B + e_{\mathfrak{p}} + e_{\mathfrak{q}}} u^{r-|\mathfrak{p}|} v^{|\mathfrak{q}|}$$

$$= \sum_{B \in \mathcal{B}(M\backslash 0)} \mathrm{Hilb}_B(M\backslash 0) \sum_{\mathfrak{p} \subseteq B} \sum_{\mathfrak{q} \subseteq [n] \backslash B} \mathbf{t}^{e_B + e_{\mathfrak{p}} + e_{\mathfrak{q}}} u^{r-|\mathfrak{p}|} v^{|\mathfrak{q}|}$$

$$= K\mathcal{T}_{M\backslash 0, M\backslash 0}^T(u+1, v+1).$$

A similar argument, with $\zeta = -e_0$, shows that the coefficient of t_0^2 in (7.3.3) is $K\mathcal{T}_{M/0, M/0}^T$. Finally, we apply Theorem 7.2.8 once more, this time with $\zeta = e_0$ and $L = H_{e_0=1}$. We find that the terms in (7.3.3) divisible by t_0 but not by t_0^2

sum to

$$\left(\sum_{\substack{B \in \mathcal{B}(M), \\ 0 \notin B}} \mathrm{Hilb}_B(M, M) \sum_{\mathfrak{p} \subseteq B} \sum_{\mathfrak{q} \subseteq [\widetilde{n}] \setminus B} \mathbf{t}^{e_B + e_\mathfrak{p} + e_\mathfrak{q}} u^{r - |\mathfrak{p}|} v^{|\mathfrak{q}|} \right)\Bigg|_{H_{e_0 = 1}}$$

$$= \left(\sum_{\substack{B \in \mathcal{B}(M), \\ 0 \notin B}} \mathrm{Hilb}_B(M, M) \sum_{\mathfrak{p} \subseteq B} \sum_{\mathfrak{q} \subseteq [n] \setminus B} \mathbf{t}^{e_B + e_\mathfrak{p} + e_\mathfrak{q}} (1 + t_0 v) u^{r - |\mathfrak{p}|} v^{|\mathfrak{q}|} \right)\Bigg|_{H_{e_0 = 1}}$$

$$= \sum_{\substack{B \in \mathcal{B}(M), \\ 0 \notin B}} \mathrm{Hilb}(\mathrm{Cone}_B(P(M, M)) \cap H_{e_0 = 1}) \sum_{\mathfrak{p} \subseteq B} \sum_{\mathfrak{q} \subseteq [n] \setminus B} \mathbf{t}^{e_B + e_\mathfrak{p} + e_\mathfrak{q}} u^{r - |\mathfrak{p}|} v^{|\mathfrak{q}|}$$

$$+ t_0 \sum_{\substack{B \in \mathcal{B}(M), \\ 0 \notin B}} \mathrm{Hilb}(\mathrm{Cone}_B(P(M, M)) \cap H_{e_0 = 0}) \sum_{\mathfrak{p} \subseteq B} \sum_{\mathfrak{q} \subseteq [n] \setminus B} \mathbf{t}^{e_B + e_\mathfrak{p} + e_\mathfrak{q}} u^{r - |\mathfrak{p}|} v^{|\mathfrak{q}| + 1},$$

which by Lemma 7.3.4 is equal to

$$\sum_{\substack{B \in \mathcal{B}(M), \, I \in \mathcal{B}(M/0), \\ 0 \notin B \qquad I \subset B}} t_0 t_{B \setminus I}^{-1} \mathrm{Hilb}_{(I, B)}(M/0, M \setminus 0) \sum_{\mathfrak{p} \subseteq B} \sum_{\mathfrak{q} \subseteq [n] \setminus B} \mathbf{t}^{e_B + e_\mathfrak{p} + e_\mathfrak{q}} u^{r - |\mathfrak{p}|} v^{|\mathfrak{q}|}$$

$$+ t_0 \sum_{\substack{B \in \mathcal{B}(M), \, I \in \mathcal{B}(M/0), \\ 0 \notin B \qquad I \subset B}} \mathrm{Hilb}_{(I, B)}(M/0, M \setminus 0) \sum_{\mathfrak{p} \subseteq B} \sum_{\mathfrak{q} \subseteq [n] \setminus B} \mathbf{t}^{e_B + e_\mathfrak{p} + e_\mathfrak{q}} u^{r - |\mathfrak{p}|} v^{|\mathfrak{q}| + 1}$$

$$= t_0 \sum_{(I, B) \in \mathcal{B}(M/0, M \setminus 0)} \mathrm{Hilb}_{(I, B)}(M/0, M \setminus 0) \left(\sum_{\mathfrak{p} \subseteq B} \sum_{\mathfrak{q} \subseteq [n] \setminus B} \mathbf{t}^{e_I + e_\mathfrak{p} + e_\mathfrak{q}} (1 + t_{B/I} v) \right) u^{r - |\mathfrak{p}|} v^{|\mathfrak{q}|}$$

$$= t_0 \sum_{(I, B) \in \mathcal{B}(M/0, M \setminus 0)} \mathrm{Hilb}_{(I, B)}(M/0, M \setminus 0) \left(\sum_{\mathfrak{p} \subseteq B} \sum_{\mathfrak{q} \subseteq [n] \setminus I} \mathbf{t}^{e_I + e_\mathfrak{p} + e_\mathfrak{q}} \right) u^{r - |\mathfrak{p}|} v^{|\mathfrak{q}|}$$

$$= t_0 K \mathcal{T}^T_{M/0, M \setminus 0}(u + 1, v + 1),$$

as desired. $\qquad\square$

Remark 7.3.5. We remark that for a general flag matroid $\mathcal{F}$, the slices $\{P(\mathcal{F}) \cap H_{e_i = k}\}_{k \in \mathbb{Z}}$ need not be flag matroid base polytopes. Moreover, even when they are, we do not observe an identity like the one in Theorem 7.3.3 that expresses $K\mathcal{T}_\mathcal{F}$ in terms of the slices. For example, consider $\mathcal{F} = (U_{1,3}, U_{2,3})$. We have $K\mathcal{T}_\mathcal{F}(x, y) = x^2 y^2 + x^2 y + x y^2 + x^2 + 2xy + y^2$. In any coordinate direction, its three slices are $(U_{0,2}, U_{1,2})$, $(U_{1,2}, U_{1,2})$, and $(U_{1,2}, U_{2,2})$, whose $K\mathcal{T}$ are (respectively), $x y^2 + y^2$, $xy + x + y$, and $x^2 y + x^2$.

Remark 7.3.6. One can generalize Theorem 7.3.3 as follows. The proof is essentially identical to one given for Theorem 7.3.3. Denote by $M^\ell := (M, \ldots, M)$ (repeated ℓ times). Then we have

$$K\mathcal{T}^{\widetilde{T}}_{M^\ell} = t_0^\ell K\mathcal{T}^T_{(M/0)^\ell} + t_0^{\ell - 1} K\mathcal{T}^T_{(M/0)^{\ell-1}, M \setminus 0} + \cdots + K\mathcal{T}^T_{(M \setminus 0)^\ell}.$$

The proof is essentially identical to one given for Theorem 7.3.3.

131

7.4 Further directions

In this final section, we state some conjectures related to the flag-geometric Tutte polynomial, and mention some directions for future research.

7.4.1 The Las Vergnas Tutte polynomial

Consider the "twisted flag variety"

$$\widetilde{Fl}(1, \mathbf{r}, n-1; n) := \left\{ \begin{array}{l} \text{linear subspaces} \\ (\ell, L_1, \ldots, L_k, H) \end{array} \,\middle|\, \begin{array}{c} \dim \ell = 1, \, \dim H = n-1, \, \dim L_i = r_i, \\ \ell \subseteq L_k, \, L_1 \subseteq H, \, L_i \subseteq L_{i+1} \end{array} \right\}.$$

The difference with the usual flag variety $Fl(1, \mathbf{r}, n-1; n)$ is that the inclusions $\ell \subseteq L_1$ and $L_k \subseteq H$ have been weakened to $\ell \subseteq L_k$ and $L_1 \subseteq H$.

The twisted flag variety fits in a diagram

$$
\begin{array}{ccc}
 & \widetilde{Fl}(1, \mathbf{r}, n-1; n) & \\
 & {}^{\widetilde{\pi}_{\mathbf{r}}} \swarrow \qquad \searrow {}^{\widetilde{\pi}_{1(n-1)}} & \\
Fl(\mathbf{r}; n) & & \mathbb{P}^{n-1} \times \mathbb{P}^{n-1}.
\end{array}
\tag{7.4.1}
$$

We can apply our usual construction to this diagram to obtain another generalization of the Tutte polynomial. Surprisingly, for two-step flag matroids, this has a nice combinatorial interpretation.

Definition 7.4.1 ([LV75]). Let $\mathcal{F} = (M_1, M_2)$ be a two-step flag matroid on a ground set $[n]$. We define the *Las Vergnas Tutte polynomial* of (M_1, M_2) to be

$$LVT_{\mathcal{F}}(x, y, z) := \sum_{S \subseteq [n]} (x-1)^{r_1 - r_1(S)} (y-1)^{|S| - r_2(S)} z^{r_2 - r_2(S) - (r_1 - r_1(S))}.$$

Theorem 7.4.2. *With the notations as above, we have*

$$LVT_{\mathcal{F}}(\alpha - 1, \beta - 1, w) = \sum_{m} (\widetilde{\pi}_{(n-1)1})_* \widetilde{\pi}_{\mathbf{r}}^* \Big(y(\mathcal{F})[\mathcal{O}(0,1)][\bigwedge^{m}(\mathcal{S}_2 / \mathcal{S}_1)] \Big) w^m$$

as elements in $K^0((\mathbb{P}^{n-1})^{\vee} \times \mathbb{P}^{n-1})[w]$.

In [DES20], the Las Vergnas Tutte polynomial is discussed in more detail, including a proof of Theorem 7.4.2.

7.4.2 g and h polynomial for flag matroids

For a matroid M, Speyer introduced in [Spe09] a polynomial $g_M(t) \in \mathbb{Q}[t]$ and a close cousin $h_M(t) \in \mathbb{Q}[t]$, which is related to $g_M(t)$ by $h_M(t) = (-1)^c g_M(-t)$ where c is the number of connected components of M. A K-theoretic interpretation of the polynomial h_M was given in [FS12].

Theorem 7.4.3 ([FS12, Theorem 6.1 & Theorem 6.5]). *Let M be a matroid of rank r on $[n]$ without loops or coloops. Let $\pi_r, \pi_{(n-1)1}, \alpha, \beta$ be as in (6.2.9). Then the polynomial h_M is the (unique) univariate polynomial of degree at most $n-1$ such that*

$$(\pi_{(n-1)1})_* \pi_r^* \big(y(M) \big) = h_M(\alpha + \beta - \alpha\beta).$$

For a flag matroid $\mathcal{F}$ on $[n]$, this motivates us to consider $(\pi_{(n-1)1})_* \pi_{\mathbf{r}}^* \big(y(\mathcal{F}) \big)$, where the maps are as in the flag-geometric construction (6.2.10). By Proposition 7.1.1, this is equal to

$$\sum_{p,q} \chi\Big(y(\mathcal{F})[\bigwedge^p \mathcal{S}_k][\bigwedge^q \mathcal{Q}_1^\vee] \Big)(\alpha - 1)^p (\beta - 1)^q.$$

Let us consider its torus-equivariant version

$$\sum_{p,q} \chi^T \Big(y(\mathcal{F})^T [\bigwedge^p \mathcal{S}_k]^T [\bigwedge^q \mathcal{Q}_1^\vee]^T \Big) u^p v^q$$

where u and v are formal variables. We show that this is a polynomial in uv, which thereby establishes that $(\pi_{(n-1)1})_* \pi_{\mathbf{r}}^* \big(y(\mathcal{F}) \big)$ is a polynomial in $\alpha + \beta - \alpha\beta$ (since the substitution $u = \alpha - 1, v = \beta - 1$ yields $1 - uv = \alpha + \beta - \alpha\beta$).

Lemma 7.4.4 (cf. [FS12, Lemma 6.2]). *Let $\mathcal{F} = (M_1, \ldots, M_k)$ be a flag matroid on $[n]$, and suppose every constituent of $\mathcal{F}$ is both loopless and coopless. Then*

$$\sum_{p,q} \chi^T \Big(y(\mathcal{F})^T [\bigwedge^p \mathcal{S}_k]^T [\bigwedge^q \mathcal{Q}_1^\vee]^T \Big) u^p v^q \in \mathbb{Q}[u, v]$$

is a polynomial in $\mathbb{Q}[uv]$.

We remark that the condition about a flag matroid $\mathcal{F} = (M_1, \ldots, M_k)$ being loopless or coopless depends only on M_1 or M_k (respectively). Note that if $\ell \in [n]$ is a loop in M_i then it is a loop in M_{i-1} also. Dually, if $\ell \in [n]$ is a coloop in M_i then it is a coloop in M_{i+1} also. Hence, the flag matroid $\mathcal{F}$ is loopless (coopless) if and only if M_1 has no loops (M_k has no coloops).

Proof. Once more by (6.1.5), we get

$$\sum_{p,q} \chi^T \Big(y(\mathcal{F})^T [\bigwedge^p \mathcal{S}_k]^T [\bigwedge^q \mathcal{Q}_1^\vee]^T \Big) u^p v^q = \sum_{F \in \mathcal{F}} \mathrm{Hilb}_F(\mathcal{F})) \sum_{\mathbf{p} \subseteq B_k} \sum_{\mathbf{q} \subseteq [n] \setminus B_1} t^{-e_{\mathbf{p}} + e_{\mathbf{q}}} u^{|\mathbf{p}|} v^{|\mathbf{q}|}.$$

Fix $|\mathbf{p}| = i, |\mathbf{q}| = j$, and consider the sum

$$\varphi_{ij} = \sum_{F \in \mathcal{F}} \mathrm{Hilb}_F(\mathcal{F})) \sum_{\substack{\mathbf{p} \in B_k, \, \mathbf{q} \in [n] \setminus B_1, \\ |\mathbf{p}| = i \qquad |\mathbf{q}| = j}} t^{-e_{\mathbf{p}} + e_{\mathbf{q}}}. \tag{7.4.2}$$

We need show that φ_{ij} is zero if $i \neq j$. Let P be the convex hull of $\{-e_{\mathfrak{p}} + e_{\mathfrak{q}}\}$ appearing in the summation (7.4.2). Note that P is contained in the intersection of $H_{e_{[n]}=j-i}$ and the cube $\{x \in \mathbb{R}^n \mid -1 \leq x_\ell \leq 1 \ \forall \ell \in [n]\}$. By Theorem 7.2.8 (in the form of Remark 7.2.9), it thus suffices to show that $\varphi_{ij}|_{H_{e_\ell=-1}} = 0$ and $\varphi_{ij}|_{H_{e_\ell=1}} = 0$ for all $\ell \in [n]$.

Let us now fix any $\ell \in [n]$. As none of the constituents have coloops (and in particular ℓ is not a coloop in M_k), the intersection $P(\mathcal{F}) \cap H_{e_\ell=0}$ is a non-empty face of $P(\mathcal{F})$ minimizing in the e_ℓ direction, consisting of bases $F = (B_1, \ldots B_k)$ such that $\ell \notin B_k$. Thus, we have that $\mathrm{Cone}_F(\mathcal{F}) \in \mathcal{P}_n^{e_\ell}$ if and only if $\ell \notin B_k$, and by Theorem 7.2.8 with $\zeta = e_\ell$ we have

$$\varphi_{ij}|_{H_{e_\ell=-1}} = \sum_{\substack{F:\ell \notin B_k}} \sum_{\substack{\mathfrak{p} \in B_k, \\ |\mathfrak{p}|=i}} \sum_{\substack{\mathfrak{q} \in [n] \setminus B_1, \\ |\mathfrak{q}|=j}} \mathrm{Hilb}((-e_{\mathfrak{p}} + e_{\mathfrak{q}} + \mathrm{Cone}_F(\mathcal{F}))|_{H_{e_\ell=-1}}).$$

But since $\ell \notin B_k$ implies $\ell \notin \mathfrak{p}$, every cone $-e_{\mathfrak{p}} + e_{\mathfrak{q}} + \mathrm{Cone}_F(\mathcal{F})$ occurring in the sum above will have vertex v with $v_\ell > -1$. Moreover, we have $\mathrm{Cone}_F(\mathcal{F}) \in \mathcal{P}_n^\zeta$ for such cones, and hence we get $\varphi_{ij}|_{H_{e_k=-1}} = 0$. A similar argument with $\zeta = -e_k$, noting that ℓ is not a loop in M_1, shows that $\varphi_{ij}|_{H_{e_k=1}} = 0$. $\qquad\square$

We thus make the following definition that generalizes the polynomial h_M of a matroid M to the setting of flag matroids. It is well-defined by Lemma 7.4.4.

Definition 7.4.5. Let $\mathcal{F} = (M_1, \ldots, M_k)$ be a flag matroid $[n]$ such that every constituent of $\mathcal{F}$ is both loopless and coloopless. Let $\pi_{(n-1)1}, \pi_{\mathbf{r}}, \alpha, \beta$ be as in Section 7.1.1. Then the polynomial $h_\mathcal{F}$ is defined as the (unique) univariate polynomial of degree at most $n-1$ such that

$$(\pi_{(n-1)1})_* \pi_{\mathbf{r}}^*\left(y(\mathcal{F})\right) = h_\mathcal{F}(\alpha + \beta - \alpha\beta).$$

Although one may also consider a similar construction via the "Las Vergnas" diagram (7.4.1), the analogue of Lemma 7.4.4 fails in this case.

7.4.3 Open problems

For matroids, the *characteristic polynomial* (also called *chromatic polynomial*, as it generalizes the chromatic polynomial of a graph) is defined by

$$\chi_M(q) = (-1)^{r(M)} T_M(1-q, 0).$$

In 2015, Adiprasito, Huh and Katz proved the following conjecture by Rota-Heron-Welsh.

Theorem 7.4.6 ([AHK18]). *Let $w_i(M)$ be the absolute value of the coefficient of $q^{r(M)-i}$ in the characteristic polynomial of M. Then the sequence $w_i(M)$ is log-concave:*

$$w_{i-1}(M)w_{i+1}(M) \leq w_i(M)^2 \ \text{for all } 1 \leq i < r(M).$$

Since we now have a definition for the Tutte polynomial of a flag matroid, we can define the *characteristic polynomial* of a rank $\mathbf{r}$ flag matroid $\mathcal{F}$ by

$$\chi_{\mathcal{F}}(q) = (-1)^{rk} K\mathcal{T}_{\mathcal{F}}(1-q, 0).$$

Note that $K\mathcal{T}_{\mathcal{F}}(x, 0) = 0$ whenever M_1 has a loop.

Conjecture 7.4.7. *Theorem 7.4.6 holds for the characteristic polynomial of an arbitrary flag matroid.*

In Examples 6.2.14 and 6.2.15, the characteristic polynomials are $-q^2 + 2q - 1$ and $4q^3 - 14q^2 + 16q - 6$, respectively. Thus, we see that Conjecture 7.4.7 holds for these examples. We can also ask for an analogue of Theorem 7.4.6 for the Las Vergnas Tutte polynomial.

Definition 7.4.8. For a flag matroid $\mathcal{F} = (M_1, M_2)$, define its *beta polynomial* $\beta_{\mathcal{F}}(q)$ by
$$\beta_{\mathcal{F}}(q) := (-1)^{r_2 - r_1} LV\mathcal{T}_{\mathcal{F}}(0, 0, -q).$$

Note that when $\mathcal{F} = (U_{0,n}, M)$, it follows from $LV\mathcal{T}_{\mathcal{F}}(x, y, z) = T_M(z+1, y)$ that $\beta_{\mathcal{F}}(q) = \chi_M(q)$, the characteristic polynomial of M.

Conjecture 7.4.9. *The (absolute values of the) coefficients of* $\overline{\beta}_{M_1,M_2}(q) := \beta_{M_1,M_2}(q)/(q-1)$ *form a log-concave sequence.*

Here is a more accessible conjecture regarding the characteristic polynomial. In contrast to our combinatorial results from Section 7.3, the formula in Conjecture 7.4.10 does not seem to have a nice T-equivariant version.

Conjecture 7.4.10. *Let M be a matroid of rank r with no loops, so that $(U_{1,n}, M)$ is a flag matroid. Then $\chi_{(U_{1,n},M)}(q) = (q-1)^r$.*

As noted in Remark 6.2.16, the flag-geometric Tutte polynomial can have a negative coefficient. However, in many examples the coefficients are nonnegative. More precisely, we conjecture the following.

Conjecture 7.4.11. *If M_1 is an elementary quotient of M_2, then the flag-geometric Tutte polynomial $K\mathcal{T}_{(M_1,M_2)}$ has nonnegative coefficients.*

Flag matroids are a special class of Coxeter matroids. Hence, another possible direction of research would be.

Problem 7.4.12. Explore how our constructions and results could be generalized to arbitrary Coxeter matroids.

Finally, we could apply the construction of Section 6.2.4 to *any* subvariety of a Grassmannian (or even a flag variety), not just to torus orbits. It could be interesting to study the properties of this invariant.